The
Domestication of
Language

The Domestication of *Language*

CULTURAL EVOLUTION
and the UNIQUENESS *of*
THE HUMAN ANIMAL

Daniel Cloud

Columbia University Press New York

Columbia University Press
Publishers Since 1893
New York   Chichester, West Sussex
cup.columbia.edu

Library of Congress Cataloging-in-Publication Data
Cloud, Daniel.
The Domestication of language : cultural evolution
and the uniqueness of the human animal / Daniel Cloud.
pages cm
Includes bibliographical references and index.
ISBN 978-0-231-16792-5 (cloth : alk. paper)
ISBN 978-0-231-53828-2 (e-book)
1. Language and languages—Origin.
2. Grammar, Comparative and general—Phonology.
3. Grammar, Comparative and general—Syntax.
4. Human evolution. 5. Historical linguistics.
6. Anthropological linguistics.
I. Title.
P116.C58   2014
417'.7—d23   2014012982

FRONTISPIECE: Rembrandt van Rijn,
The Anatomy Lesson of Dr. Nicolaes Tulp, 1632.
(169.5 cm × 216.5 cm, oil on canvas, Mauritshuis, The Hague)

COVER DESIGN: *Catherine Casalino*

When I was a child I pictured our language as settled and passed down by a board of syndics, seated in grave convention along a table in the style of Rembrandt. The picture remained for a while undisturbed by the question of what language the syndics might have used in their deliberations, or by dread of vicious regress.

I suppose this picture has been entertained by many, in uncritical childhood. Many mature thinkers, certainly, have called language conventional. Many have also, in other connections, been ready with appeals to agreements that were historically never enacted. The social contract, in Hobbes' theory of government, is the outstanding example. This case is logically more respectable than the language case; the notion that government began literally in a social contract involves no vicious regress.

Not, of course, that the proponents of the social contract mean to be thus construed; they mean only that government is as if it had been thus established. But then this "as if" proposition raises the question, psychoanalytically speaking, of latent content: in just what ways is government like what an actual social contract might have given us? In the language case this question of latent content is even more urgent, and more perplexing, in that an original founding of language by overt convention is unthinkable. What is convention when there can be no thought of convening?

W. V. Quine, foreword to David Lewis, *Convention*

Contents

1. Where Do Words Come From?   1

2. The Conventions of a Human Language   37

3. The Evolution of Signals   79

4. Varieties of Biological Information   91

5. The Strange Case of the Chimpanzee   113

6. The Problem of Maladaptive Culture   143

7. The Cumulative Consequences of a Didactic Adaptation   163

8. Meaning, Interpretation, and Language Acquisition   183

9. What's Accomplished in Conversation?   203

10. Recapitulation and Moral   241

References   *251*

Index   *265*

The
Domestication of
Language

1

Where Do Words Come From?

QUINE'S PARADOX OF THE SYNDICS

How did all the various things in the world get their names?

> And out of the ground the LORD God formed every beast of the field, and every fowl of the air; and brought *them* unto Adam to see what he would call them: and whatsoever Adam called every living creature, that *was* the name thereof. (Genesis 2:19)

That's one theory. Who stands behind the lexicographers, or the teachers, and whose laws are they enforcing when they insist that we use words "correctly" rather than "incorrectly"? Perhaps all the meanings were decreed by Adam himself, the ultimate ancestor, in a single original act of baptism. And perhaps our job, as philosophers, is to re-create that original mother tongue or universal character, and rescue humankind from its long captivity in Babel.

Perhaps those things are true, but it sounds like a fairy tale. Our languages are part of our inherited culture, and it now seems wrong to think of our own cultural conventions as the decayed remnants of an ancient dialogue between the first man and God. As philosophers working in a tradition that has often been preoccupied with language, we ought to be able to come up with a better story. So it should come as no surprise that a number of philosophers, ancient and modern, have tried to account for the origins of language in a more convincing way.

Plato (1961) made one of the earliest attempts in *Cratylus*. His way of posing this question has been influential—the quotation from W. V. Quine that serves as the epigraph to this book is one echo of it, as we'll soon see. Aside from the suggestion made by Hermogenes, in the dialogue, that our language must be a system of arbitrary conventions, however, most of his tentative proposals concerning its possible answers seem to have struck modern philosophers as implausible.

But implausibility isn't always a defect. Sometimes a very alien way of looking at a problem offers us exactly the new perspective we need. Plato was not naive or foolish. In *Cratylus*, starting with very different assumptions, he made some suggestions about where the meanings of words come from that we now find surprising. Are they all just bad ideas, in spite of their source? Or could some of them be hidden treasures, insights that look unpromising simply because they're so unfamiliar?

Here I want to revive in a more contemporary form what seems superficially to be one of the least plausible of those ideas. Socrates is depicted as believing that the words of our existing languages were created by people he calls *nomothetes*, lawgivers or legislators. Although Quine called this idea childish, to Plato it apparently seemed obvious. Since neither of them actually knew where the meanings of words came from, this is purely a difference of intuition between two great philosophers, one ancient and one modern. That alone should make us curious.

This Socratic story, too, initially sounds like a fairy tale about heroic founders, like the story about Adam in Genesis. Reading the dialogue that way, however, is a mistake. It's putting the quaint medieval thing we think Plato must have been saying in place of what the text actually says. Although Socrates does talk later about the hypothetical "first legislators" who might have created the most basic elements of human languages before the beginning of recorded history, when he first introduces this idea he's obviously talking about a process of invention that was supposed to be still going on in his own society at the time of writing. Laws, after all, were still being made in Athens, so the legislators were still around.

"Can you at least say who gives us the names which we use?" Socrates asks Cratylus. "Does not the law seem to you to give us them?" (*Cratylus* 388d). These questions aren't about what might have happened long ago, any more than those following them are: "How does the legislator make names?" "To what does he look?" (*Cratylus* 389a). When he first starts talking about the nomothete, Socrates isn't telling a fairy tale. He's making an observation about the way things work in his own world, in the present. The use of language, he argues in this part of the dialogue, is a technology, like weaving and shipbuilding: "A name is an instrument of teaching and distinguishing natures, as the shuttle is of distinguishing the threads of the web" (*Cratylus* 388c). Lawgivers manufacture its tools, just as the maker of looms makes tools for the weaver. Just as it's the weaver and not the loom maker who's the proper judge of looms, he tells us that it's the teacher and not the lawgiver who's the proper judge of words.

I don't want to defend the more fabulous hypotheses about unobserved events in the prehistoric past that Plato gives us later on in the dialogue. What I do want to revive here is the idea that our existing languages are partly the product of ongoing human invention and human judgment, that particular individuals did, and still do, play a role in deciding what our language will be like that is something like the role of Quine's imagined "syndics."

How could anyone possibly believe that? The idea that human languages are invented by particular human individuals, "legislators," appears so ridiculous to us now that it's hard to pause before rejecting it even long enough to ask what the original claim was. But what was Plato really saying? Why would he have blamed the legislators, in particular?

The lexicographer wasn't yet available as a suspect. The first recognizable precursor to the modern monolingual dictionary seems to have been compiled several decades later, in Alexandria, by Philitas of Cos (Sbardella 2007). Like *Cratylus* itself, it was largely focused on clarifying the meanings of obsolete or obscure words used by Homer, though the dictionary also included technical terms and words from local dialects. So like *Cratylus*, it was a response to the rapid evolution of a natural human language, to the highly visible differences between Homeric and Attic Greek. This connection to the evolution of the Greek lexicon explains the otherwise somewhat puzzling contents of Plato's dialogue, in which a subtle philosophical inquiry into the origins and nature of language is interwoven with abstruse speculations about the sources of various Homeric and modern words. Aristotle's philosophical lexicon in the *Metaphysics*, probably written decades later than *Cratylus*, seems like a bold technical innovation, given that the members of the philosophical generation that preceded Socrates were still writing philosophy in verse.

Euclid's *Elements*, which begins with a list of formal definitions, hadn't yet been written. Mathematical definitions, in some sense of the word *definition*, may have been important to Plato, but his writings don't contain many recognizable formal mathematical definitions of a Euclidean kind. In fact, in *The Forgotten Revolution* (2004:171–202), Lucio Russo argues that they're a slightly later invention.

One type of definition that would have been familiar to Plato, however, is the legal definition. It's hard to have laws or a constitution, as ancient Athens did, without explicitly defining lots of terms like *citizen*, *property*, and *murder*. Formal contracts rely on explicit definitions of their terms or on tacit and customary definitions that can be made explicit. In court, lawyers argue about the proper interpretation of those terms, about whether or not they're actually in the possible world specified by one of the contract's clauses. Since the founders of Greek cities often wrote their first constitutions, and great constitutional reformers like Solon explicitly redefined terms like *citizen* as part of their reforms, it must have seemed reasonable to project this deliberate, rational, explicit, socially central activity all the way back past Homer to unknown original definers existing at some time in the distant past. The overall hypothesis would then be simple continuity, that the present is a good clue to the past, that things probably got the way they are now as a result of the sorts of deliberative processes we still see going on all around us all the time. In this kind of story, the meanings of words are settled in the course of our ongoing efforts to settle the meanings of words.

At the very beginning of this extrapolated process, Plato saw more or less the same paradox that Quine pointed out in the opening epigraph when he spoke about syndics and vicious regress. To Plato, though, the paradox was a genuine puzzle about our prehistory, and not a reason to reject the whole idea of human agency. Here's his version of the problem:

> SOCRATES: Let us return to the point from which we digressed. You were saying, you remember, that he who gave names must have known the things which he named. Are you still of that opinion?
>
> CRATYLUS: I am.
>
> SOCRATES: And would you say that the giver of the first names also had a knowledge of the things which he named?
>
> CRATYLUS: I should.
>
> SOCRATES: But how could he have learned or discovered things from names if the primitive names were not yet given? For if we are correct in our view, the only way of learning or discovering things is either to discover names for ourselves or to learn them from others.
>
> CRATYLUS: I think there is a good deal in what you say.
>
> SOCRATES: But if things are only to be known through names, how can we suppose that the givers of names had knowledge, or were legislators, before there were names at all, and therefore before they could have known them?   (*Cratylus* 438a)

Apparently the assumption is that we obtain our most fundamental knowledge—our knowledge of the necessary or analytic truths that Platonists supposed were the sole contents of mathematics or any other genuine science, our knowledge of what's "true by definition"—at least partly from the meanings of words, from our language itself. The original nomothetes, the very first legislators or, more literally, "rule givers," who supposedly created the most basic elements of our original languages, must have had another method of knowing those first truths, since they incorporated them in their creations, but what was it? Sensory information alone, without any logical rules or preexisting categories to use in classifying and interpreting it, seems inadequate.

The worry appears to be more or less the same as the one about a vicious regress that Quine expressed; although if Quine's "syndics" are analogues of Plato's nomothetes, Quine was more skeptical of their actual existence. In Plato's paradox of the first nomothetes and Quine's paradox of the syndics, both philosophers are in fact pointing to a philosophical puzzle we still face in almost the same form.

Anticipating his own later work, Ludwig Wittgenstein ([1922] 1998) asserted in *Tractatus Logico-Philosophicus* that "all propositions of our colloquial language are actually, just as they are, logically completely in order" (5.5563). When he said this, he was saying something that appears to be true—but why should that be so if our language was handed down to us from people who knew much less about the world than we do? How did ordinary language ever get into such good order in the first place? Who could have arranged it that way without having a language that was already in perfectly good order to use in figuring out how to do it? When did they do that, and what did doing it consist of? Were they doing something different from what we do now? Was there some extra human activity at some point in the past that resulted in words acquiring the meanings they have at present? Why and when did we stop doing whatever it was? Or are we still doing whatever activity has had those consequences over time, even now, without realizing it? John L. Austin (1956) claimed that ordinary language "embodies . . . the inherited experience and acumen of many generations of men" (11). How did their experience and acumen become embodied there? Plato's willingness to focus on this paradox and, in the end, admit that he had no real solution for it is arguably better than the modern tendency to simply ignore it or push it off into the unobservable past.

Without the somewhat fabulous notion of "original legislators," which Plato used partly to motivate his own version of Quine's paradox of the syndics, his suggestion presumably would just be that all the way back to the unknown beginnings of human language, people have sometimes disagreed about what words like *citizen*, *murder*, and *delivery* mean, or should be taken to mean, and that by settling these disputes, prestigious or powerful third parties like Solon have, over time, created communal consensus about their correct meanings. The story also seems to include some further process in which teachers, as their users, make judgments about the usefulness of the words that the legislators have given us in this way, about whether a new definition of a word is better than what came before it.

Even if that isn't true—even if that sort of adjudication is never necessary or is so exceptional as to be unimportant, even if people never argue about what was meant by something that was said or what a word really means, or even if those billions of iterated arguments have had no cumulative consequences at all—the idea isn't obviously absurd or genuinely childish, so our own immediate impulse to reject it is a bit puzzling. Why should Plato's theory about where meanings come from strike modern people as ludicrous?

The problem, I think, is that in a modern context, Plato's fable about the origins of meanings quickly runs into another, much more thoroughly entrenched fable. When the imagination of most modern people is directed to the same unknown origins of the same natural human languages, it produces a very different picture. We naturally gravitate toward some version of behaviorism: a distant ancestor must have habitually made a particular sound when he carried out some action, and perhaps from this association, his peers came to regard that as a name for the action, or else an imitation of a sound gradually became a way of referring to the thing that made that sound. Then things continued to happen in more or less the same perfectly and completely accidental and unconscious way until finally we ended up with the very complex human languages we have today.

But this, too, is recognizably just an ancient philosophical myth about the origins of language in the distant and unobservable past, a myth that Brian Skyrms (2010:5) attributes to Democritus. It's the myth that became a modern folk belief, perhaps because it fits so well with David Hume's idea that we learn about the world mostly by unconsciously keeping track of recurring associations. That doesn't mean it's the truth about the actual evolutionary history of the human animal. The part of the theory that tells us how we get from the caveman's crude protolanguage of groans and grunts to the full complexity of a modern human language doesn't seem to have many details. The classical fable that instantly sounds right to us could be just as fabulous as the one that instantly strikes us as wrong.

To my own surprise, I've come to believe that there's an element of truth in the apparently less plausible Platonic story that's easy to miss, one that seems to be almost completely obscured by the paradox that both Quine and Plato have described. It isn't that our languages were deliberately invented by particular groups of people, legislators or syndics in the formal sense of those words, sitting around particular tables, at particular times in the past. It seems to me that they're more like our dogs, our wolfhounds and sheepdogs and dachshunds, our retrievers and pointers and greyhounds. We didn't invent them exactly, but our ancestors did repeatedly make deliberate, more or less rational choices in the process that made them what they are today, choices among a long series of slightly incrementally different variants, unconsciously shaping the dogs into precisely what their human breeders needed them to be.

What most people missed about this activity for most of the time they were engaged in it was that its results accumulate in the powerful way they do. A long series of small, seemingly inconsequential choices among accidentally occurring variations produced big, consequential changes over time. The animals were optimized, mostly inadvertently and without much coordination by their breeders, as specialized tools for particular human activities. The same thing could be true of the more voluntary aspects of our daily use of language. We may have simply overlooked the way their results accumulate.

The case of sheepdogs may offer a particularly good analogy. Many of the choices humans have made about which sheepdogs to keep and breed from have been made in the course of their efforts to manage sheep, so the choices were mostly made in passing, while doing something else, but the existence of that other, more immediate, objective doesn't mean they still weren't choices about dogs.

The lecturer in the painting used as the frontispiece for this book, Rembrandt van Rijn's *The Anatomy Lesson of Dr. Nicolaes Tulp*, is trying to manage his society's accumulated knowledge of human anatomy, trying to get the members of his audience to improve their culturally transmitted ideas about the newly discovered lymphatic system. (There are other theories about what he might have been talking about, but it makes no difference to the argument.) He's helping a piece of culture—our new knowledge about the lymphatic system and everything needed to make that new knowledge comprehensible—propagate through a professional community, surgeons. He's using language to do it, though, choosing his words carefully, and the members of his audience are likely to be picking up the way he uses terms even as they struggle to grasp the anatomical knowledge he's using them to convey.

Although Dr. Tulp may be choosing his words in passing while trying to do something else, that doesn't mean he's not still choosing them, and choosing them carefully. He's renaming important parts of their professional world, and the surgeons who are familiar with only the old names, or their old senses and referents, appear to be attending closely to this nomothetic act, like anxious shepherds thinking of buying a new breed of dog to herd their sheep.

If that's a good analogy, if words really are something like sheepdogs, then we need a philosophical theory of where their meanings come from that's very different from anything we currently have. They aren't *just* conventions, though that's one of the things they appear to be, and they aren't *just* tags attached to objects long ago, though that, too, is one of the things they can be. They're also carefully curated pieces of culture subject to cultural evolution, like every other kind of human culture.

Of course, there's nothing new about the idea that the meanings of names are passed down to us from previous generations. That's just Saul Kripke's (1980) historical-chain-of-transmission theory of reference (Burgess 2013:28–33). There's nothing new about the idea that speakers may introduce variations in meaning as the name passes down the chain (Evans 1973; Kripke 1980:163). There's also nothing very novel about the general idea that conventions can evolve. H. Peyton Young's "The Evolution of Conventions" (1993) is a classic of evolutionary game theory. Skyrms (2010) wrote a book about the evolution of signaling conventions, in particular. Austin (1956) spoke of words in ordinary language as having "stood up to the long test of survival of the fittest"(8). What may be less familiar is the idea that the evolution of the conventions of our language might involve what Charles Darwin called "artificial selection."

Daniel Dennett (2009a) has already suggested that domestication might be a good model for the evolution of technical terms, "anchored by systematic definitions fixed by convention and reproduced in the young by deliberate instruction, rehearsal, and memorization" (4), though he expressed some doubts about the model's applicability to ordinary language. But we play a direct, and at least partly conscious, role in selecting the versions of *all* our words that will be heard or read by other people and therefore might be reproduced. This suggests to me that domestication is the only model of the evolution of words in general that can possibly fit the facts.

If the meanings of words are part of our cultural inheritance, if we learn them from those around us, and if they are "conventional," in David Lewis's ([1969] 2002) specific sense of the word *convention* (which I will explain in chapter 2, since it's different from the sense that someone like Young [1993] might attach to the word), then it seems to follow logically that they must be thought of as domesticated, in Darwin's ([1859] 2009:36–48) sense of the word *domesticated*, that we have no choice. Something becomes domesticated when we acquire a veto over its reproduction. The farmer chooses which seed to plant and rejects the others. The shepherd picks which dogs to breed from, or keep, and rejects the others. For something to be a convention, in Lewis's ([1969] 2002) sense, as opposed to, for example, an unconsciously imitated mannerism, or an absolute practical necessity, it must be possible, in principle, for us to choose to reject it in favor of some "almost equally good" alternative (68–76). Darwinian domestication occurs when we're put into a position to reject some candidates for inclusion in a gene pool on the basis of our own human preferences, perhaps in pursuit of purely local and temporary goals, like catching rabbits or herding sheep, and thereby increase the contribution of other alternatives.

Self-interested participation in conventions is another requirement of Lewis's theory. Otherwise, they become obligations requiring sacrifice, which is a different phenomenon. We rationally choose among the available conventions regarding the meanings of the words we might use when we speak, choosing some precise sense in which to use them or choosing to use entirely different terms. If we do this in the same self-interested way that we rationally accept or reject greyhounds or racehorses for breeding and they're subsequently passed on to others for acceptance or rejection in the same way, over and over, how can they *not* follow the same general evolutionary pattern as other domesticated replicators?

If we don't rationally accept or reject conventions regarding the meanings and associations of words, what was Lewis trying to do by writing *Convention*? Wasn't he trying to persuade us to adopt an amended and clarified convention regarding the meaning of the word *convention*, as the book's title suggests? Was he trying to persuade us to change behavior—our way of interpreting and using that word—about which we have no choice?

DOMESTICATION AS A HYPOTHESIS

To argue in favor of such a strange hypothesis seems a bit quixotic. Certainly words are closely associated with their human users, but are there really any good reasons to speak in terms of "domestication"? Isn't it a bit self-inflating to assign ourselves that amount of control over their evolution? Surely if this way of looking at human language made sense, everyone would already know that.

I can understand these doubts. Implausibility may not always be a defect, but that doesn't mean it's usually a virtue, and this idea *is* a little strange. But the reasons for entertaining the hypothesis strike me as compelling. I can only hope that I can argue for my point of view convincingly enough to make the skeptics feel that trying to understand the argument was worth the effort, even if they can't find a way to agree with it.

What are those reasons? We sometimes speak about "the evolution of language" as if it is some discrete event that happened in the distant past. That seems wrong to me. If our languages have evolved at all, it seems to me that they must still be evolving now. But in the present, we see ourselves making conscious, rational decisions about which words to use and what to say we mean by them. Like Darwin, I think the default assumption should be that the present is often a good clue to the past, that things probably became what they are now as a result of the sorts of processes still going on all around us today, happening over and over again, for a very long time. The only way to reconcile the imperative of explanatory continuity with the evidence of everyday experience is through a story about linguistic and cultural evolution that at least partly depends on a mechanism of selection involving repeated human choices.

The conscious or unconscious artificial selection of domesticated organisms is precisely that kind of mechanism, as Dennett (2009a, 2009b) pointed out in talking about the evolution of words and music. It's a mechanism that Darwin described very clearly.

The prescience of any human selector is necessarily limited, so the consequences of their actions may go well beyond those they intended, but that's no reason to deny domesticators the limited amount of myopic rationality they actually do display. They choose among the options their environment offers, culling the mutations that displease them and conserving the pleasing ones. We seem to do the same thing with words, and their associations, and senses, and references, accepting some of the subtle and small incremental innovations we encounter and rejecting others. We adopt some of the ways we hear words being used (often during childhood but sometimes as adolescents or adults) and pass them on to others. Other ways of using words fail to please or impress us and soon drop out of our personal idiolect (an idiolect is an individual's idiosyncratic version of a language) or never enter it. Some new variants, unattractive or humanly useless ones, are culled while others proliferate through the idiolects of human populations.

The association seems beneficial to both us and, in terms of their reproductive interests, the words. We appear to have a veto over the further propagation of a word, or a particular meaning for a word, through the simple expedient of never repeating it, never using it or never using it in a certain sense, never making use of certain associations, or never using it to refer to certain things that's a bit like the veto we have over the reproduction of a domestic plant or animal. Who ever speaks now of a *humor*, meaning "bile" or "choler"? Who ever uses the word *condescending* in the way it was used in *Pride and Prejudice*, as a term of praise? Who now uses *approve* to mean "verify," or *hoist* to mean "blown up," as Shakespeare did in *Hamlet*? What stays is only what we ourselves keep and repeatedly use, because only our repeated use of a word in some particular sense can create opportunities for others to learn and adopt our way of using it.

The choice of which word to use and which sense to use it in is tautologically a "selection event." Yet it's just as tautologically an intentional choice. So some part of the evolution of our language's entrenched conventions really must be thought of in this way, as Darwinian evolution under what Darwin would have called "artificial selection."

Although Dennett deserves the credit for first pointing out this possibility, he probably wouldn't be willing to agree that domestication is the correct theory of the evolution of most words in natural human languages. Instead, he argues in favor of what he calls a "neutral framework," in which we admit the possibility of a whole spectrum of selection processes, ranging from domestication to natural selection without human intervention.

Why do we disagree, and what exactly do we disagree about? The difference may be partly one of emphasis. Dennett points out the wide range of processes that might be operating in the evolution of a human language. I'm interested in investigating one particular process. If any words, or any part of language, really can be thought of as domesticated to any degree, it seems to me that the discussion is only begun, and not ended, by that insight. It appears to raise as many questions as it answers, so there's still some room for differing opinions about the scope and details of the phenomenon.

The difference might be partly one of emphasis, but in at least one passage in "The Cultural Evolution of Words and Other Thinking Tools" (2009a), Dennett seems to unambiguously reject domestication as a theory of the evolution of most words in ordinary, everyday language, arguing that it only really applies to terms from formalized technical languages. He introduces the idea that words might perhaps be thought of domesticated, but he then seems to drastically restrict its range of application.

Dennett begins "The Cultural Evolution of Words and Other Thinking Tools" by talking about the association between people and words as often being a "mutualism," a mutually beneficial association. The relationship between honeybees and the various kinds of flowers they pollinate is mutualistic, since the bees get nectar and the flowers are pollinated.

Domestication is a kind of mutualism, though not all mutualisms can be thought of as domestications. As Darwin ([1859] 2009) used the term, *domestication* refers to humans' cultivation of animals, plants, and other organisms. In this sense, the critical feature of domestication is the human role in choosing which individuals will have offspring and which will not. The farmer chooses which ears of corn he will take the seeds for the next generation of maize plants from, and the hunter chooses which of his dogs he'll breed from and which of the litters he'll keep. Ants do something very similar with aphids, herding them, protecting their eggs, even sometimes taking an egg with them when they start a new colony. Because of its obvious similarity to the domestication of animals by humans, we refer to what the ants do as "domestication" even if we're not sure whether they are choosing which eggs to conserve on the basis of any qualities in the aphids that lay them.

By analogy, any mutually beneficial interaction that becomes obligatory for one or both parties can be called domestication in the broadest possible sense of the word, because the obligatory character of the interaction gives one or both parties a kind of "veto" over the reproduction of the other. No fig trees means no fig wasps, so in this sense the wasp has been domesticated by the tree. The association between honeybees and particular species of flowering plant doesn't meet this additional condition, though, because the honeybee can always get nectar from some other kind of flower and the flowers pollinated by honeybees typically also have other pollinators. The title of an interesting paper— "Polydnaviruses of Parasitic Wasps: Domestication of Viruses to Act as Gene Delivery Vectors" (Burke and Strand 2012)—about an obligatory mutualism between wasps and viruses contains a typical example of this way of using the word.

After first suggesting that many words must be in mutualistic relationships with their human users, however, Dennett (2009a) warns us that not all their interactions with us may be so benign: "Our paradigmatic memes, words, would seem to be mutualists par excellence, because language is so obviously useful, but we can bear in mind the possibility that some words may, for one reason or another, flourish despite their deleterious effects on this utility" (3).

He then introduces the somewhat narrower idea of domestication, apparently using the word in Darwin's sense. But Dennett (2009a) immediately casts doubt on the applicability of this idea outside modern, specialized technical contexts:

> Although words now have plenty of self-appointed guardians, usage mavens, and lexicographers who can seldom resist the temptation to attempt to legislate on questions of meaning and pronunciation, most words—almost all words aside from coinages tied tightly to particular technical contexts—are better seen as synanthropic, like rats, mice, pigeons, cockroaches, and bedbugs, rather than domesticated. They have evolved to thrive in human company, but nobody owns them, and nobody is responsible for their welfare. The exceptions, such as "oxygen" and "nucleotide," are anchored by systematic definitions fixed by convention and reproduced in the young by deliberate instruction, rehearsal, and memorization.   (4)

There are good theoretical reasons for Dennett to worry about the possibility that from our point of view, our association with many words may be maladaptive. There seem to be sound evolutionary arguments suggesting that many of our associations with items of culture might be deleterious to our fitness. I'll discuss some of them in subsequent chapters; basically they derive from Richard Dawkins's (1976) original analogy between an item of culture, what he called a "meme," and a virus. Like viruses, items of culture are numerous and reproduce rapidly, in a pattern that's markedly different from that of their human hosts. In the last few thousand years, there seems to have been a lot more cultural evolution than human biological evolution, so apparently memes evolve much more quickly than we do. Because items of culture can be passed on without their carriers passing on their genes, they have a way to escape from their human hosts even if they've somehow lowered those hosts' Darwinian fitness. The analogy with actual viruses—which also are numerous, rapidly reproducing, and rapidly evolving replicators that also have ways of escaping from their hosts even after harming their fitness—seems compelling. Like many people who study cultural evolution, Dennett is, in thinking about the evolution of words, explicitly adopting this memetic perspective. That makes his views on the subject of cultural evolution much closer to the reigning orthodoxy than the view that I will defend here.

Dennett draws two major conclusions about the evolution of words from his memetic theoretical approach. In general, he argues, it must be true that

1. Excellently designed cultural entities may, like highly efficient viruses, have no intelligent design at all in their ancestry.
2. Memes, like viruses and other symbionts, have their own fitness. Those that flourish will be those that better secure their own reproduction, whether or not they do this by enhancing the reproductive success of their hosts by mutualist means. (Dennett 2009a:3–4)

Words, he suggests, are no exception.

While Dennett does mention the alternative of domestication, much of his argument in favor of this classical memetic perspective on the evolution of words is directed instead against what he calls "traditional wisdom." But Darwin's model of domestication also conflicts with these two stipulations. Darwinian domesticators do make more or less intelligent choices about which seeds to plant or which dogs to breed from, so there is some sense in which the ear of corn or the dachshund or the greyhound is partly a product of "intelligent design": "The key is man's power of accumulative selection: nature gives successive variations; man adds them up in certain directions useful to him. In this sense he may be said to have made for himself useful breeds" (Darwin [1859] 2009:37). Human populations have grown dramatically as a direct consequence of the domestication of plants and animals, so those plants and animals have flourished at least partly by enhancing our reproductive success. It seems less likely that maize or cattle would have been domesticated as successfully as they have been if every human population that began to domesticate them immediately started to shrink.

Whatever our traditional wisdom on the subject of the evolution of words may be, my intention here is to defend the hypothesis that words in general are domesticated, in Darwin's specific sense of the word *domesticated*, as a Darwinian theoretical alternative to the sort of memetic perspective that Dennett is advocating. Dennett (2009a) thinks that "almost all" (4) words are like rats or bedbugs; I'll argue that they're more like dogs or maize. To make my point of view plausible, I must find a way to persuade the reader that it isn't true that words in human languages have "no intelligent design at all in their ancestry" (3), though I'd prefer to replace the somewhat loaded term "intelligent design" with the more neutral "rational choice." I also must find a way to argue that words typically do thrive by "enhancing the reproductive success of their hosts by mutualist means" (4), that people are seldom harmed by learning a new word and often are made better off by it.

Although by arguing these things, I may be disagreeing with what now seems right to many people, neither case seems all that difficult to make. It seems obvious that we do use some intelligence in choosing our words and which of the various, perhaps only slightly different, available senses to use them in. (More than none at all, which is what the phrase "no intelligent design *at all*" seems to imply.) It's hard to think of a case in which learning a new word would really make someone worse off or even less likely to have children, less "fit." Learning how to use the terminology associated with some technology for birth control might do that, but then the real cause is the technology and the innate human desire to control our own reproduction (Hrdy 2000), not its associated vocabulary. Even if this sort of thing sometimes does happen, it seems difficult to argue convincingly that this has historically been the typical situation in the human relationship with language.

Certainly Dennett and the other advocates of the "memetic perspective" must exercise at least some rational choice in deciding to use words like *memetic* instead of some alternative, and they presumably do their best to make sure that they use words like *domesticated* in the best way they know of. If I started using these words in some nonstandard or obsolete sense, I would probably be criticized for having failed to choose, in an intelligent, honest, and rational way, a standard and useful version to employ.

Dennett (2009a) does make an exception for words "such as 'oxygen' and 'nucleotide,' . . . anchored by systematic definitions fixed by convention and reproduced in the young by deliberate instruction, rehearsal, and memorization" (4). So perhaps his careful choice of the word *memetic* or our care to use the word *domestication* properly doesn't count as evidence that language, in general, is domesticated. This might just be evidence that modern technical languages are domesticated.

But I am inclined to doubt that we need two completely different theories of the evolution of words, one for technical terms in modern languages and the other for everything else. If the process by which technical language evolves really is as different from the evolution of ordinary language as Dennett seems to be suggesting, then we need some further explanation about how the two processes became so different, about when and how we adopted such a radical new technology for managing certain parts of our language.

I address the historical origins of modern technical languages at the end of chapter 9. It was clearly a momentous event. I don't think that it was the initial domestication of language, however. Instead, it seems to me to have merely been a switch to a more advanced set of techniques and institutions for cultivating it, like the modern switch to more advanced and self-conscious ways of breeding animals and plants that enabled Darwin to discuss the process with such accomplished experts. He certainly didn't argue that plants and animals became truly domesticated only with the emergence of these experts and their explicit and conscious programs of breeding for certain traits.

If language became domesticated only when it became technical, in some specifically modern sense involving systematic definitions and rehearsal, that would have been a recent and dramatic change in human biology, the beginning of a qualitatively different mutualistic association with language. Until then, as far as we were concerned, all our words would have been like rats or bedbugs. After this change, some of the words would have become, for some of us at least, like maize and cattle. Some of us would suddenly have become capable of using words, rather than being used by them. This, however, strikes me as seeing ourselves as too special. It appears to make the modern, highly educated person into a whole different type of creature from the rest of humanity, a creature who freely chooses to use domesticated words while all other human populations are just unavoidably infested by synanthropic ones.

It seems to me that evolution, and not revolution, is a better theory of the origin of modern technical languages, because to me what Euclid did in the *Elements* doesn't look completely different from what Socrates was doing in the dialogues, or from what we do when we explain how to play chess, or tie a bowline knot, or make a roux, or use the Levallois technique for knapping flint. It strikes me as an incremental modification of very common and ancient human practices, not a way of making our language into something entirely new and different.

Dennett says that nobody owns words and nobody is responsible for them. It seems to me that I do own the words in my own personal idiolect, though I don't own the whole species each of them belongs to, because I can freely dispose of them and have control of their fate. I did quite a lot of work to acquire them in their current forms. I also can't help thinking that we actually are responsible for the words we use and for the senses we use them in, though perhaps not for their "welfare," whatever that means for a word. We certainly are partly responsible for their Darwinian "fitness," for how many opportunities they have to leave copies of themselves behind in other minds, and for whether or not they're likely to succeed. We're the only available suspects. Things don't apply names to themselves.

Saul Kripke (1980) and Hilary Putnam (1975) contended that *people* name things, which appears to make at least some of us responsible for their names. The painting used as the frontispiece for this book initially looks like a picture of such a dubbing ceremony. It's as if Dr. Tulp is naming the various parts of the lymphatic system one by one, just as Adam is supposed to have named the animals. There's something a little odd about thinking of this as a picture of that sort of event, though, because presumably this was not the first time any audience had ever heard the new terms and had their application pointed out. There must have been a number of such conversations as the new knowledge percolated through the professional community of surgeons, fixing the reference of the new terms for a whole series of audiences. Each time, for each new audience, it must have seemed as if the things were being named, or renamed, for the very first time, right in front of their eyes, an impression deliberately reinforced in the lecture depicted by the presence of the corpse itself, by the direct display of the objects being named. If it seemed to his audience that Tulp was applying names irresponsibly or haphazardly, they certainly would have held him responsible, asking questions that began "How can you possibly say . . ." or "Do you really think it makes sense to say . . . ?"

We can learn things from works of art, and in this case, I think, we can learn something important about naming ceremonies. I think we learn that these ceremonies must be recapitulated and repeated many times to be effective. Dr. Tulp's demonstration couldn't possibly have been the only one. Things don't stay named over long periods of time unless we periodically reiterate or recapitulate the naming ceremony for new audiences, as we do in teaching or pointing out locations on a map or writing a textbook. This is the empty set, and *this* means "or"; this is Dartmouth; this is I-25; this is a carbon atom; this is the integral sign; this is the carburetor; this is the superior *vena cava*; this is a bowline knot; this is flint, and this is chert; this is a coyote track; this is the scat of a wolverine; this is a yew tree; this is a roux; this is a mouse, and this is its house. The historical chain of transmission is at least partly a historical chain of recapitulations of the original act of bestowing the name. Often we describe the item as we give its name: "You see, the leaves are long and narrow."

When we do this, even though we're usually just trying to conform to convention, sometimes we must introduce subtle or obvious novelties, by chance or inspiration, which may or may not prosper. From the spectrum of possible meanings for each word, we're choosing specific ones and advocating in their favor, saying that they're the correct meanings and the others are incorrect. "To make a real roux, on the other hand, you have to . . ." Of course, the chefs presumably think they're saying what a roux is, not what the word *roux* means, but the information they present about the first thing is perfectly adequate to tell us the second, especially if they give it in the presence of the object itself. "A wolf has bigger feet; these are the tracks of a coyote." Again, saying what a wolf and a coyote are like also tells us something about what the words *wolf* and *coyote* mean. We take responsibility for the "welfare" of words, I think, or of particular, very slightly different meanings for words, at least to the extent that we participate in anything like this, any naming or renaming or reiteration of naming ceremonies, any "teaching and distinguishing natures," to borrow Plato's description. It's a process that occurs in many different ways.

When I look closely at the particular examples Dennett suggests as analogies for the human relationship with ordinary words, I find it hard to agree that they're genuine parallels. I can't make myself believe that most words are like rats or mice or roaches or bedbugs because I don't see how it's possible that the typical word thrives by hiding from human attention. Even feral pigeons are skittish around people and tend to nest in abandoned buildings or other inaccessible places. Mutual avoidance doesn't really strike me as the essence of the human relationship with language.

(Although Dennett starts out by talking about mutualisms, all the examples he gives seem to be examples of human commensals: animals that live alongside us and benefit from our presence without benefiting us very much in return. It's difficult to think of an undomesticated organism that's in a genuinely mutualistic relationship with modern humans. It might seem that the microbes living in our intestines are an example, but they live there because our immune system lets them. Strains of *E. coli* that aren't as well behaved provoke a severe immune response. If the human brain is to words what our immune system is to our internal microorganisms, that would seem like an argument in favor of, not against, domestication because in that case, we clearly do have a veto.)

The cuckoo seems as if it might be an exception to the rule that parasites are selected to evade the host's notice. Aren't cuckoo chicks, nest parasites that eject all the other eggs from the host's nest, louder and more colorful than the host's own chicks? But the host's only hope of saving its own brood is detecting the cuckoo's egg while it's still an egg. Once it has hatched, sometime during the day when they are away foraging, and the invader has rolled all the host's own eggs out of the nest, it will be too late to save this season's brood. Consequently, the hosts are mainly selected for their ability to detect eggs. The eggs of nest parasites are often carefully camouflaged and can be very hard to tell from those of the host, if the association is an old one. The parasite becomes noticeable and starts manipulating the brain and senses of the hosts only later, when the brood is already lost and selection on the host for the ability to detect the intruder is relaxed. The host can't save the season at that point, so they will not be passing on any genes this year no matter what they do. The parasite still can pass on its genes, however, so selection on it is not relaxed at all, making this portion of the arms race rather easy for it to win (Dawkins and Krebs 1979). The cuckoo is inconspicuous when it's still an egg, when it needs to be inconspicuous. Is there some similar cryptic phase in the life history of words, a time when they disguise themselves as something else?

The crucial question is whether we make meaningful choices among the slightly varying meanings that we encounter. These choices don't have to involve any intent to improve the language; we just have to have preferences and act on them. We just have to feel that some of the ways of using words we encounter aren't reasonable enough or honest enough or pleasant enough to persuade us to adopt them ourselves. Darwin's ([1859] 2009) whole point about artificial selection was that it was largely inadvertent:

> [F]or our purpose, a form of Selection, which may be called Unconscious, and which results from every one trying to possess and breed from the best individual animals, is more important. Thus, a man who intends keeping pointers naturally tries to get as good dogs as he can, and afterwards breeds from his own best dogs, but he has no wish or expectation of permanently altering the breed.

Nevertheless we may infer that this process, continued during centuries, would improve and modify any breed, in the same way as Bakewell, Collins, &c., by this very same process, only carried on more methodically, did greatly modify, even during their lifetimes, the forms and qualities of their cattle.   (40)

Official "mavens" were never supposed to be required.

If there exist savages so barbarous as never to think of the inherited character of the offspring of their domestic animals, yet any one animal particularly useful to them, for any special purpose, would be carefully preserved during famines and other accidents, to which savages are so liable, and such choice animals would thus generally leave more offspring than the inferior ones; so that in this case there would be a kind of unconscious selection going on.   (42)

Aristotle apparently said something similar about proverbs in his lost *On Philosophy*, suggesting that they're rare survivals from ancient philosophical traditions, saved by their conciseness and cleverness when everything else was lost in the catastrophes that periodically engulf humanity (Synes, Calvit. Enc. 22 85c, in Ross 1952:79). In both cases, our ancestors are supposed to have simply used the best variants they could find, preserving them through all disasters and thus passing them on to subsequent generations, inadvertently modifying their inheritance by preferentially saving the ones they happened to find most useful.

To claim that we don't exercise at least this amount of volition in choosing our words—that we don't remember the useful ones and forget the useless ones, that we don't generally use words in ways we think will work well and avoid using them in ways we think will work badly, that we make no effort to be persuasive or speak well or sound "cultivated," or at least like an adult when we choose them or to describe things accurately or interpret descriptions faithfully—seems too strong to me. When we say, "That's not the word I would have chosen," I think we're referring to an actual voluntary process of choosing our words in conversation. When we say, "That's not what I meant," I think we mean that we were trying to use a word in a particular sense, a sense that wasn't the one that appears to have been apprehended. Not only is choice being exercised, but in a case like this, explanations of the choice and why it was the best one can often be given.

This doesn't mean we understand the full consequences of our locally optimizing choices. The naive domesticator makes more or less locally rational choices among the available alternatives, pursuing short-term and local ends in ways that can have randomly fortuitous or perverse consequences over the long term. There were plenty of fortuitous accidents and unexpected side effects in the domestication of organisms like dogs or maize or tobacco, and lots of more or less rational choices about what to preserve in each generation and what to cull. Really irrational human choices are often just noise. Inventing the lightbulb, or even slightly improving it, is more likely to matter on a large scale than driving home drunk—but since human societies are incomprehensibly complex, most of the consequences of even the most rational choice will probably be unintended. It's still a locally optimizing choice; we've still repeatedly chosen the best word we could find and used it in the most appropriate sense we were aware of, with whatever evolutionary consequences may flow from those facts.

I also find it difficult to agree with Dennett that ordinary words are merely "synanthropic" because I think that this theory fails to explain the most significant features of the phenomenon of human language as we see it every day. It doesn't explain Wittgenstein's ([1922] 1998) observation that natural human languages are "logically completely in order" (5.5563). It doesn't explain why they're so well adapted to the uses that we make of them. This is something that a theory of their evolution ought to explain. Words seem to be adapted to our convenience, just as other tools are. In fact, our words sometimes seem to be wiser than we are. The theory that they're merely synanthropic—like rats, roaches, mice, pigeons, and bedbugs—doesn't explain what Nelson Goodman ([1955] 1983) would have called the "projectibility" of the "entrenched predicates" of natural languages, the extent to which they're useful for forming thoughts that allow us to make accurate predictions about the world. That is adaptation; in fact, it's very elaborate adaptation—and if it isn't at least partly adaptation to our needs, I don't know what would be.

Would individual humans typically be just as fit, or even more fit, if they were never exposed to any words, as they might be if they were never exposed to any bedbugs or rats? From actual cases, we know that a human without any early experience of human language or communication, of words or signs or communicative gestures of any kind, is a very vulnerable and helpless creature who, in the wild parts of Africa or in any other natural environment, would probably not survive for long. ("Secrets of the Wild Child," an episode of *Nova* from 1997, is a fascinating but very sad record of one such case.) We appear to be as dependent on words, or other closely related forms of communication such as sign language, as termites are on their cellulose-digesting intestinal bacteria.

So in the end, I'm more inclined to agree with Dennett when he defends his hypothesis that at least some words are domesticated than I am when he expresses caution about the applicability of that hypothesis to ordinary language. The memetic approach and domestication strike me as two competing theories about the evolution of the meanings of words, not as complementary and compatible theories, and I think there are good reasons to prefer the hypothesis of domestication. Because of their utility and their complete dependence on human choices for their propagation, words in general seem to me to be more like dogs and horses and maize, selected to be useful or pleasing to humans, than they are like bedbugs, rats, and other merely synanthropic creatures, selected to evade our notice and surreptitiously pilfer our treasures. We use words. We don't use bedbugs. Like other domesticated replicators, they're our tools or pets, not parasites or scavengers in our trash, even though they're tools we may often use with perverse and unforeseen consequences.

In fact, since cultural evolution and the evolution of human languages are processes that must long predate the domestication of animals and plants, it now seems to me that it might be more appropriate to say that Darwin himself identified only a special, recent case of a much older and more general human adaptation, one that's so much a part of our nature that it has a tendency to escape our notice, just as water might easily escape the notice of a fish. We apparently have had the ability to cultivate and incrementally improve various evolving cultural practices and tools, perhaps including words and their senses, by attending to quality, culling out items that are of little use, and saving the best examples, for a very long time. Only in fairly recent times have we figured out how to extend this process of cultivation to animals and plants. In human terms, how different is an ear of corn from an obsidian spearhead or from the word *corn*? All these things are tools; all of them are curated items of culture. If we think of them as subject to cultural evolution, don't we have to think of all the memes involved as domesticated? Considering the amount of effort involved in learning to make an obsidian spearhead, isn't the spearhead just as dependent for its reproduction on the investment of human labor and the locally optimizing choices of individual humans as any maize plant? Isn't it very likely that our ancestors deliberately *selected* the best spearhead they could find to copy?

What makes humans' domestication of other living things different from ants' domestication of aphids is our ability to do it with a great variety of organisms, rather than just one kind, and to domesticate animals and plants we've only recently encountered. The de novo domestications of new creatures, yaks, and tomatoes, and *Penicillium roqueforti* are unmistakably cases of cultural innovation, so it makes perfect sense to assimilate this novel and recent form of cultural change to other processes of cultural evolution that have been taking place in human societies for much longer periods of time. Once this has been done, however, the details begin to seem suggestive as comments on their new context; they appear to tell us something about cultural evolution in general. If an ear of corn is like a tool or a culturally transmitted skill, then why isn't a tool or a skill like an ear of corn?

Some people might object that there's really no such thing as "domestication" or "artificial selection" in the first place, because humans are part of nature and we and our plant and animal partners evolve by the same processes as everything else. Surely what that word is referring to is just coevolution between two organisms in an ecological relationship of mutual dependence?

And that's exactly what it is. But the domestication of animals, plants, and other organisms by humans is a very rare and unusual kind of coevolution. In all but a few cases, only the organism on one side of the relationship has changed as a consequence of the interaction. Some humans have acquired the ability to digest the milk of cows, even as adults, but our various breeds of cattle have undergone even more complex and visible changes. Domesticated animals and plants can change dramatically, but their human domesticators change only slightly or, more commonly, not at all.

No other organisms have so many mutualistic relationships with so many kinds of animals and plants and yeasts and molds and bacteria and fungi, with new partnerships still being formed all the time. The skills needed to keep particular kinds of domestic organism—farming practices, ways of making cheese, and so on—seem to be transmitted and preserved in much the same way as other skills and tools and ideas, other items of culture. But to the extent that it involves cultural evolution, the human domestication of animals and plants and microbes certainly is very different from ordinary, purely biological coevolution, or even from the domestication of aphids by ants.

Domestication, as Darwin understood it, also seems to involve its own distinctive form of selection. "Artificial selection" is somewhat like sexual selection in that in both cases a system of sensory organs and a brain are directly involved in the selection process. Darwin himself pointed out this analogy between artificial and sexual selection in *The Descent of Man* ([1871] 2004). In discussing sexual selection, he noted that "this latter kind of selection is closely analogous to that which man unintentionally, yet effectually, brings to bear on his domesticated productions, when he continues for a long time choosing the most pleasing or useful individuals, without any wish to modify the breed" (684–85).

In both cases, attention must be paid if reproduction is to succeed. The seed must be chosen and planted by the gardener; the peahen must notice the peacock's tail. The same thing is true of words: if they're to be repeated, they must be heard and attended to, so there's also a system of sensory organs and a brain directly involved in their replication.

At a slightly earlier point in the same book, Darwin ([1871] 2004) also made an analogy between the evolution of domesticated organisms under artificial selection and the evolution of our own human cultural practices for self-adornment— that is, for painting and scarring and tattooing our faces, shaping our children's skulls, binding their feet, and putting plates in their lower lips and large disks in their ears:

> In the fashions of our own dress we see exactly the same principle and the same desire to carry every point to an extreme; we exhibit, also, the same spirit of emulation. . . . The same principle comes

largely into play in the art of selection; and we can thus understand, as I have elsewhere explained, the wonderful development
of all the races of animals and plants which are kept merely for
ornament. Fanciers always wish each character to be somewhat
increased; they do not admire a medium standard; they certainly
do not desire any great and abrupt change in the character of their
breeds; they admire solely what they are accustomed to behold, but
they ardently desire to see each characteristic feature a little more
developed.   (651)

A long incremental series of slight exaggerations, one after the other, he
suggested, must have led in the end to the large lip plate or the custom
of severely deforming an infant's skull, just as they led to the Pekingese
dog or the Fantail pigeon. Darwin, too, saw a close analogy between
this sort of cultural evolution and domestication, as well as an analogy
between domestication itself and sexual selection. He made more or less
the same comparison between artificial selection and the cultural evolution of human self-adornments in *The Variation of Animals and Plants
Under Domestication* (1868:1:214–18, 2:240).

Would the implied similarity between the domestication of items
of culture, as humans allegedly do it, and sexual selection, as peacocks
and peahens do it, have any observable consequences? In the preced-
ing quotation, echoing his earlier remarks in *The Variation of Animals
and Plants Under Domestication*, Darwin observed that the breeders
of domesticated animals and plants, like animals that are the targets
of sexual displays, seem to always prefer slightly exaggerated versions
of characteristic and unusual features. Perhaps the extreme beauty of
many of the products of human culture, the poetic richness of human
languages, and the beauty of the displays and decorations that birds use
in sexual selection are similar and are all recognizably "beauty" to a creature with senses and a brain like ours, for a reason. Their similarity can
be seen as evidence of an underlying similarity in the selective processes
that have produced them. Beauty may help a tool or pattern or picture
or song or ritual or ornament or poem or another item of culture create
more copies of itself, as it does for a peacock or a bird of paradise or a
domesticated strain of tulips. We might take all this beauty as evidence
of a long history of endlessly iterated human choices.

Think of the song lines of ancient Australia, which R. M. W. Dixon (1996), a linguist who studies that continent's many languages, argues may be more than ten thousand years old. If the songs were not beautiful or the stories they contained were not interesting, they might not have lasted and served hundreds of human generations as tools for navigation. Some of the words of the old songs must have been carried along with them. The *Iliad*, too, is a piece of oral culture, set up to be easy to recite from memory. The archaic but beautiful Greek names and other old words that it contains were part of what motivated Plato to write *Cratylus*.

In the evolutionary history of humans, true modernity is associated with the practice of decorating tools and making ornaments, of using ocher to adorn ourselves, of scratching regular patterns into the ostrich shells we use to carry water, of finding and keeping seashells to use as beads. Does this new tendency, in the last hundred thousand years or so, for conserved cultural artifacts to become more and more beautiful reflect a human preference for the eye-catching and the exaggerated? If so, it would be clear evidence of the domestication of human culture, in Darwin's sense of the word *domestication*.

Have our languages and our cultures also domesticated us, at least in the broader Burke-and-Strand (2012) sense of the word *domestication*? Is the association obligatory on both sides? Here we have to be careful. For their reproduction, their wasps are as dependent on *particular* viruses as the viruses are on the wasps. A human, in contrast, must be exposed to *some* culture and *some* language in order to turn into a fully developed adult, but various alternatives are often available to us. Groups of humans can even make up their own language, as has happened quite recently with Nicaraguan Sign Language (Kegl 1994), though they might need to be in contact with other language-using humans to accomplish that. No part of human culture can survive without humans, but individual humans can survive without many parts of many human cultures, just as they can survive without tulips or maize, so any particular mutualism between a human and an item of culture is obligatory in only one direction.

We have to have *some* language and *some* culture, just as a honeybee has to interact with *some* flower—but we can choose which items of culture, or which version of each item, we'll associate with. It's the existence of these choices that gives us a true Darwinian veto, a veto that we can exercise in a way other than by simply being absent, so it's precisely this one-sidedness that gives domestication by humans its unique character. For us, our associations with words are mutualisms, just as they are from the point of view of the words. It's just that from the point of view of any particular word, or version of a word, it's an obligate mutualism, whereas from our point of view, the association is "facultative," or optional. It's this asymmetry that makes us the more powerful partner in the interaction, and it's partly the asymmetry in power that makes human cultural evolution so different in its effects and its rapidity from mutualistic coevolution in general.

SEMANTIC EVOLUTION

What are we human breeders of human dialects supposed to be choosing between?

The process of learning syntax, learning to speak "grammatically," is largely unconscious. This suggests to me that syntax must be a set of what Lewis ([1969] 2002) called "mannerisms" (in contrast to "conventions"), unconscious imitations of those around us that are not subject to rational rejection. Although these syntactic mannerisms must evolve, since the process of learning to imitate them is mostly unconscious and automatic, it isn't clear that we can think of them as domesticated in Darwin's narrow sense of the word. (Since they obviously can't survive without humans, they do seem to be "domesticated" in the broader, Burke-and-Strand [2012] sense.) If we want to explain the evolution of syntax, we actually do have to try to reconstruct what might have happened to our ancestors in the distant past, and we do have to think in terms of a process that usually involves little conscious human choice in the present. Similar remarks could be made about the sounds of speech, about clicks and tones and other aspects of phonology. These are the kinds of things many linguists study, but they aren't what I am talking about here.

Semantics in a human population—the public understandings of the meanings of words shared by the speakers of its language or languages—is very different. The meanings of words, in the Lewisian theoretical framework that I'll explain in chapter 2, aren't unconscious verbal mannerisms; they're rationally rejectable conventions.

This suggests that acceptable grammar and the tacit or explicit definitions or senses of words must evolve by very different processes. I think that's true. Words jump horizontally between languages easily and routinely and are subject to conscious borrowing (Dixon 1997:19–27). The very existence of pidgin languages shows that new words are easier for human adults to consciously and intentionally learn than new grammar is.

A language may also borrow syntax or sounds from other languages, partly as a result of the very common phenomenon of bilingualism, but the process is less conscious and seems less routine. Words can be introduced gradually, one by one, over a long period of time, whereas changes in grammar or pronunciation are often fairly abrupt, systematic, and far-reaching. In a purely oral culture, under the influence of a neighboring language, a whole new system of cases may emerge as a unit, over a period that can sometimes be as short as a few decades (Dixon 1997:54–56). This seems to suggest that something like Noam Chomsky's "language instinct" must be partly responsible for the details of these changes. It suggests that there is a finite set of grammars that the human brain can easily adapt to. Under certain circumstances, often involving the conflicting evidence provided by bilingualism, groups of people with no written language will, unconsciously and in unison, converge on a new member of this restricted set.

Change in the lexicon, or in the meanings attached to items in the lexicon, seems different. It appears to be much more under our conscious control. As Dixon (1997) says, "People tend to think of their language in terms of its dictionary (the open classes of noun, verb, adjective). They identify it with the lexicon and attempt to control this as an indicator of ethnic identity. They are not in the same way aware of grammatical categories (or forms) and are scarcely aware of changes that may take place in this area" (24).

No innate instinct tells us what a word like *tenrec* or *convention* is supposed to mean. No instinct informs us that we must call polenta "grits" in order to be a true southerner. It's the evolution of the *conventions* of the language of a human population, our conventions concerning what words should be taken as meaning, that's the main topic of this book. I will deal with the evolution of our involuntary verbal mannerisms only in passing, because what I want to explore is what we can learn about human nature from the idea of domestication.

The process of semantic evolution, once you know what to look for, seems much easier to observe than the evolution of syntax, because it appears to be going on all around us, all the time, and involves actions that are much more under our conscious control. What does it look like? The picture used as the frontispiece for this book can continue to serve as an illustration. In it, Dr. Tulp is explaining a fine distinction or correcting some misconception about the inner parts of a man's arm, exhibited before his audience of surgeons to demonstrate his point, telling them that they must not mistake the x for the y, as so many do, or showing them some surprising features of his skillfully dissected subject.

The doctor is engaged in a highly demanding cognitive act, one very characteristic of humans: advocating one explanation of what something is and is not, which includes, in passing, comments on what various names do and do not imply. The members of his audience of surgeons also are engaged in very human, very cognitively demanding acts, interpreting his supposedly authoritative explanation and deciding whether to accept it in preference to their previous suppositions, in the process perhaps picking up some new vocabulary or new ways of using old vocabulary. There is a set of unambiguous referents, a deliberately circumscribed and carefully displayed set of things being spoken about—the fine white threads and nodes of the lymphatic system inside the corpse's arm—whose names are being given and the proper associations of each explained. They're being distinguished from one another and woven into the existing web of associations in the group's shared language.

Presumably much of the definition and redefinition of words in demonstrations is indirect and oblique. Nevertheless, enough of it takes place to convey their meanings from generation to generation and person to person. In the same way that the idea of a primitive social contract is actually shorthand for a process in which people either willingly acquiesce to the political arrangements in which they currently find themselves, or else refuse to or must be compelled to, Quine's syndics strike me as stand-ins for speakers and audiences in the present, with the listeners, in situations like the one depicted in the painting, either inwardly or outwardly acquiescing or objecting to the senses, referents, and associations that the speakers would like to attach to their words. In this kind of conversation, apparent meanings are explained and critiqued, and supposed connotations are directly compared with the real world to see if they actually match it. Someone is showing someone something, and for the sake of clarification, they're conversing during the demonstration. How many times has this happened in human history? How many times does it happen every day, all around the world?

Inherited meanings are passing through two distinct filters in this sort of discussion, the speaker's choice of what to say, of whether and how to modify or reinforce his meanings, and the listener's choice of whether to believe and modify her own way of speaking, of whom to listen to, and what to remember and repeat later. In conversation, the speaker and the listener repeatedly switch roles, so each may act as an element in both parts of this complex filter.

Much of our absorption of language isn't like this, of course. Much of it is far more casual, and often we're paying much less attention when it happens. We may pick up the meanings of many words in a fairly thoughtless process of imitation. Most of us, most of the time, may just be trying to use the word in the same way as the other people around us. But it's one of the surprising implications of the formal models of cultural evolution I'll look at later in this book that a few scattered critical learners, a few careful explainers and alert audiences, or a few arguing legislators or good teachers, sprinkled throughout a human population, can have dramatic effects on the long-term dynamics of the associated population of cultural behaviors, even if most people are just imitating other people most of the time.

This intelligent filtration doesn't have to be organized or coordinated. It can be distributed in space and time, and it doesn't have to be particularly efficient. But according to the models, something like this must have been going on somewhere in order for human culture to have existed for very long in the first place. A population in which all the imitators are completely uncritical apparently can't sustain a human kind of cumulative culture at all.

In a "domestication" theory of human culture, that's not surprising. In the absence of natural or sexual selection in a more conventional form, populations of domestic animals would quickly deteriorate without constant culling by attentive humans, whether those humans are thoughtful or merely irritable, because there's nothing else to prevent the accumulation of deleterious mutations.

In the modern world, this kind of massively parallel Socratic selection goes on all the time, in a way not matched by any equally widespread and vigorous effort to discuss fine points of grammar. The world is full of people who show other people things and explain or demonstrate what they mean by terms or who accuse one another of twisting words. Some audiences are credulous, others less so. There's no reason to believe that this is a recent development: Could our early ancestors really have learned to make and manage a fire without ever being shown what is and what is not "firewood" or "kindling"?

If not, then the continuous chain of demonstrations and explanations reaches back over a very long period of time. The dissector in Rembrandt's painting could just as easily be demonstrating some sophisticated method of knapping flint or poisoning an arrow or attaching a stone blade to a haft or pointing out the bent grass stalks where the python was sleeping yesterday. All the facial expressions would be more or less the same, though the costumes and postures probably would be different. Although the painting captures a particular moment in modern history, it's also possible to see something timeless and universal in it, an accurate depiction of characteristically human behaviors.

Because so much of this activity of getting meanings straight is under our conscious control, we tend to assume we understand it, and in some senses of the word *understand*, we do. But there's also a sense in which the activity itself and its aggregate effects in a human population over time are as mysterious as any other human social behavior. We know exactly how to do it, but that doesn't mean we know exactly what it is we're doing when we do, exactly what the full consequences of our actions will be. The cognitive transparency of our own intentions and the fact that we often reach our own little local goals as a result of them can blind us to the true complexity of our behavior and its consequences over time.

Because intentional behavior isn't arbitrary or rationally opaque, we're tempted to abandon our analysis of it when we understand the end that's being sought by the human actors. That's the "reason" for the action; why seek any other? The value of domestication as a model is that it gives us a way to see beyond that distracting mirror and into what's on the other side. It lets us see what comes next, what might result from the iteration of millions of such little goal-seeking exercises.

Caught in the illusion of rational transparency like an insect caught in amber, all through our lives we continue consciously and laboriously learning, arguing about, and teaching the nuances and details of the rich and densely interconnected meanings of words, seeing only the immediate effect on ourselves and those with whom we're directly interacting, supposing the whole time that our own arguments and investigations and explanations and doubts leave those meanings unchanged for others. After all, we were never really trying to change them, we were just trying to get them right or to ascertain them or elucidate them or make them explicit or prove that somebody else was getting them wrong. Nevertheless, our actions are bound to have cumulative consequences.

Most of these actions must simply help stabilize our language and keep it from deteriorating into nonsense, but they also seem to gradually alter parts of it. Our shared system of imaginable or describable possible worlds slowly changes as we discuss its details. Over slightly longer periods of time—since Chaucer or even Shakespeare—dramatic changes in vocabulary and meaning are still taking place in a natural language like English, just as they were in Homeric Greek.

Although we're Darwinians with respect to everything but ourselves, when it comes to human culture and history, we see our own conscious intentions too clearly and their complex unintended consequences too dimly. To escape from this illusion, I think we have to try to see our own choices and descriptions and definitions and discussions as natural phenomena, as components in a complex evolutionary process that has been going on for a very long time. The missing, hidden, mysterious, long-lost original process by which words got their meanings isn't really missing at all; we just need to recognize it for what it is, to learn to see in more Darwinian terms behaviors we've known about for a very long time.

It's this intuition about the active role of human judgment in the creation, maintenance, and expansion of our languages and cultures that I want to discuss in this book, using the generally evolutionary approach to the problem of the origins and maintenance of natural human languages and human cultures that's forced on us now by the known facts of anthropology and natural history (Boyd and Richerson 1985, 2005; Boyd and Silk 2012; Dawkins 1976; Dennett 1995; Malthus [1798] 1983;

Richerson and Boyd 2004). Dennett's brilliant analogy with Darwin's model of unconscious artificial selection appears to me to be the key to reconciling these two things.

In speaking of the "evolution of language," there are actually two different problems we might be referring to. One project is understanding how the meanings of words change over time. A related but distinct project with the same name is that of determining how our ancestors became capable of using language in the first place. In one case, what is evolving is the language itself. In the other, it is the human beings who use the language.

While these two problems are different, they're closely connected, and I've found that it isn't possible to say anything about the first question without looking carefully at the second. We can't understand what language is and how it changes without thinking about where it came from, about how humans acquired the ability to manage something like this. My story about domestication relies on the Darwinian assumption of continuity between the processes we can observe in the present and the mechanisms of evolution in the distant past, so according to its own internal logic, it should go all the way back to the beginning. If it's any good, it ought to shed light on the evolutionary origins of human language in general, by helping us understand what sort of adaptation it is. I will argue that the hypothesis is a fruitful one in that respect, that some things that are puzzling about the initial evolution of the human capability to use language and have a culture seem to become less puzzling in the context of a model involving domestication.

Thinking about evolutionary origins has forced me to think about things like the nature of biological information in general, the differences between humans and chimpanzees, and the evolutionary role of human intelligence, and to look at some simple models of the evolution of other aspects of human culture. I won't attempt to survey or review all the various literatures I have had to touch on in doing this. In most cases, others could do, and generally already have done, a better job of that. Instead, I'll confine myself to discussing the relatively few sources I need to carry out my chosen philosophical task. I make no claim that I'm covering everything important, or even the most important things, in each field.

Before I start talking about all those subjects, however, I must discuss a few of the ideas of two other philosophers, more recent than Plato, who've also thought about where the meanings of words come from. The ideas about the nature and origins of language that can be found in David Lewis's *Convention* and Brian Skyrms's *Signals* will be my contemporary philosophical point of departure. Since I'm claiming that Lewis's account of the conventions of a human language commits us to a theory of their evolution that involves Darwinian domestication, I must begin by describing enough of that account to allow readers to judge whether I'm right.

2

The Conventions
of a Human
Language

SKEPTICISM ABOUT CONVENTIONS

In order to say very much about David Lewis's *Convention* ([1969] 2002), it's necessary to first say something about the philosophical problem that he was trying to solve there, and its relationship to the one that I am investigating here. Providing that explanation requires a brief description of a small part of the philosophical background against which the book was written.

Lewis was explicitly responding to W. V. Quine's "Truth by Convention" ([1936] 2004a) and the other things Quine had written on the so-called conventions of our language and their relationship to analytic truth. In turn, in "Truth by Convention," Quine himself was merely extending an argument first made by Bertrand Russell ([1921] 2010), so it seems reasonable to start the discussion by briefly looking at Russell's ideas.

In general, Russell seems to have been a fairly orthodox adherent of Democritus's theory of the origins and subsequent development of human languages. He acknowledged the tradition of regarding the meanings of words as in some sense conventional, but he felt that was true only "with great limitations." Anticipating Quine, he suggested that this idea is basically just a myth, like the social contract. Certainly, Russell ([1921] 2010) admitted, scientific and technical terms can be adopted by deliberate convention, but

> the basis of language is not conventional, either from the point of view of the individual, or that of the community. A child learning to speak is learning habits and associations which are just as much determined by the environment as the habit of expecting dogs to bark and cocks to crow. The community that speaks a language has learnt it, and modified it by processes almost all of which are not deliberate, but the result of causes operating according to more or less ascertainable laws.   (138)

Why did he think that the kinds of explicit discussions people have in the sciences were a bad model? He insisted that understanding a word does not and cannot consist of being able to "say what it means." In fact, that sort of explanation has nothing to do with the real meanings of words. Instead, a person understands the meaning of a word when "(a) suitable circumstances make him use it, (b) the hearing of it causes suitable behavior in him" (Russell [1921] 2010:143). According to Russell, "There is no more reason why a person who uses a word correctly should be able to tell what it means than there is why a planet which is moving correctly should know Kepler's laws" (144). (But if some planets could tell us why they go around the sun in elliptical orbits, wouldn't we be entitled to seek an explanation of this capability?)

In this story, we seem to learn the correct behavioral responses to particular words through the same sort of mental habit that leads us to assume that similar causes will have similar effects. We all are constantly imitating imitations of imitations of imitations of imitations, and that's all there is to human language. There are no public conventions about the meanings of words. There are no conventional definitions. There are no correct or incorrect explanations. There is just this individual mental habit, and the behavior it produces.

Where the meanings of words came from in the first place was a problem that Russell ([1921] 2010) pushed off into the unknowable past: "The association of words with their meanings must have grown up by some natural process, though at present the nature of the process is unknown" (138). At some point long ago, we did something, but it isn't what people are doing in the sciences now, nor is it something ordinary people are still authorized to do for themselves.

Quine's contribution was taking this behaviorist theory of meaning to its logical conclusion. In "Truth by Convention" ([1936] 2004a) and "Two Dogmas of Empiricism" ([1951] 2004b), he extended Russell's doubts about the existence of linguistic conventions to analytic truths, claims that are supposed to be "true by definition." If there are no real conventional definitions behind our use of language, if that isn't the right way to explain our linguistic behavior, then, Quine reasoned, nothing is ever true by definition, except in a few artificial and extraordinary situations. In this theory of language, it isn't necessarily true that all bachelors are unmarried, because our individual linguistic behavior can easily differ in ways that make it false as far as some of us are concerned. In fact, there are no necessary or analytic truths, only beliefs that we're more or less willing to abandon.

The quotation from Quine's foreword to *Convention* that serves as an epigraph to this book preserves some of the flavor of the argument of "Truth by Convention." That argument revolves around the idea that we tend to think of analytic truths as resulting directly from our own ancient conventions about the meanings of words. But, Quine argued, the meanings behind the analytic truths of logic can't be purely conventional in any straightforward sense, because the "conventions" of logic are required even to state them. In fact, we can't truly understand them unless we've already accepted them. What's the use of pretending we ever had a chance to make a deliberate and explicit convention among ourselves about which language we'd speak and what its words would mean that somehow made them true? Why not just say that we seem to accept them as true and seem very reluctant to abandon them, even when faced with what might appear to be contrary evidence, and leave it at that?

He considered the possibility that our own attempts to define words might help establish their public meanings, but Quine ([1951] 2004b) rejected the idea because it seemed to him to reflect confusion about the purpose of the lexicographer's activity: "Clearly this would be to put the cart before the horse. The lexicographer is an empirical scientist, whose business is the recording of antecedent facts; and if he glosses 'bachelor' as unmarried man, it is because of his belief that there is a relation of synonymy between those forms, implicit in general or preferred usage prior to his own work" (35). Does it matter what the lexicographer believes he's doing? Quine seems to have felt it was important, but Darwin's theory of domestication might make us doubt that it is.

The idea that we share conventions about the meanings of words doesn't survive in Quine's account of language, but the notion of "truth" had better survive, since the acceptance of some sentences as true in certain circumstances is all that gives them whatever meaning they may still have. Without conventional definitions, however, the notion that some sentences are "true by definition"—true because of the very meanings of the words they contain—can't survive either. Somehow the things we say generally must be unambiguously true or false, even though nothing we say is ever necessarily true, true by definition. "Snow is white" must be unambiguously true, even in the absence of any shared, public conventions about what "snow" is or what counts as "white."

Quine did try to give us a sort of substitute for necessary or analytic truths. In "Two Dogmas of Empiricism," he observes that some of our beliefs are more "central" and some are more "peripheral," more and less subject to revision in the face of new data. In the complete absence of any public conventions about this, however, nothing seems to prevent individuals from assigning to their beliefs idiosyncratic degrees of centrality or peripherality, and nothing seems to force us to revise our beliefs similarly on the basis of similar evidence. One of us may regard the claim that bachelors are unmarried as central and the claim that John is a bachelor as peripheral, while another regards the claim that John is a bachelor as central and the claim that bachelors are unmarried as peripheral and easily abandoned. Finding out that John is married will produce different conclusions about his bachelorhood for each of us.

Without public conventions about the meanings of words, we might worry that each individual is free to understand every sentence he hears or speaks in his own idiosyncratic way and to reinterpret or twist the words in it in any way he pleases. It seems that each person can have his or her own strange idiolect, which can drift arbitrarily far apart. One, it would appear, is as good as another. In the absence of public conventions, it also isn't clear how our language can be repeatedly transmitted from generation to generation without our individual ways of using words being completely randomized over time by noisy transmission. But then why should we be confident that we can identify "the" truth conditions of particular sentences or say which synonymies do and don't "exist"? If we can't, then it seems they don't mean anything in particular.

These consequences, or something like them, must have seemed so unpalatable to Lewis and his philosophical generation that he, at least, found himself forced to question Russell's original rejection of the whole idea of public linguistic conventions and, with it, the validity of Quine's paradox of the syndics. Is it really true that there's no way a group of people can ever *tacitly* arrive at a convention about something? In the real, material world, couldn't one of the syndics just raise an eyebrow and say "Table . . . ?" in a way that obviously proposed this as a name for the table they were all seated at, and couldn't the other syndics just glance at one another and nod in unison? Wouldn't that establish a convention? If all that is possible, then maybe the rather extravagant philosophical consequences that Quine drew from his modernized version of Plato's paradox of the first nomothetes, his doubts about the very existence of necessary or analytic truths, the meanings of individual words, and so forth were, in the end, not warranted. The prospect of escaping from such odd conclusions by such an apparently simple move must have seemed very attractive.

For the attempted escape from Quine's paradox of the syndics to succeed, what the syndics actually did in thus tacitly agreeing to adopt the convention that the word *table* would stand for tables, and not chairs, would have to be thoroughly analyzed. The stroke of genius behind Lewis's *Convention* was his realization that Thomas Schelling's (1966) new ideas about "focal points" in "coordination games" could serve as the basis for a theory of how that might work.

Lewis's strategy was to try to understand conventions in general, and linguistic conventions in particular, as basically being conserved equilibriums in coordination games, built around established precedents. This allows the conventions to be tacit, while making our choice to follow them still a fully rational one. We often choose to follow the practices we see others following because it makes practical sense for us to do so, but we don't need a written account of the whole system of rules or any overarching rationale to make that rational choice. We just have to be able to figure out what's expected of us in particular situations well enough to be able to produce the right behavior most of the time. Those other things are still useful and may be acquired at a later point, but it's the mere ability and willingness to conform to existing conventions that is the minimum requirement for successful play. In trying to conform, we may accidentally introduce variations, so our frequent inability to give a general and precise description of the regularities that our behavior must display in order to conform to the reigning conventions might even be, from an evolutionary point of view, a creative force in their elaboration and sequential modification.

COORDINATION GAMES AND PRECEDENTS

What is a "coordination game"? It's sometimes useful to divide games into three general classes. Pure noncooperative games are games like poker and "chicken," in which the players have no significant common interests. These games are relatively easy to analyze because they often have a unique equilibrium: everyone simply does his or her worst, knowing that all the others will do the same.

What's the opposite of a noncooperative game? It seems obvious that it should be a "cooperative game," but in fact that's only partly right. There are at least two very different alternatives to noncooperative games, because there are two different things missing from a noncooperative game: there are no common interests, and there are no contracts. Although common interests are required to motivate contracts, contracts aren't a necessary condition for the existence of common interests. We may end up coordinating our actions, in a mutually beneficial interaction, either with or without the security of an enforceable contract, but these are two very different kinds of interaction.

Two people playing a zero-sum game have no reason to enter into a contract with each other, because there's no situation in which one wins without the other losing. Anything that one player is willing to agree to, the other should refuse. This situation can be altered in two different ways: either by giving the players common interests and the ability to commit to binding contracts, or by giving them only common interests. Games of the first kind are true "cooperative" games, games in which players can commit to binding contracts regarding their actions in the game before the start of play. Since there usually are lots of contracts they might adopt, corresponding to all the ways in which they could agree to divide the winnings, our questions about the likely outcome tend to become questions about the process by which the contracts are negotiated and coalitions are formed.

Games in which the players do have common interests and genuine choices to make about how to collectively pursue them, but can't commit to binding contracts, perhaps because they lack any mechanism for enforcing them, are what Lewis ([1969] 2002), following Schelling (1966), called "coordination games."

The existence of multiple possible equilibrium outcomes (multiple pure strategy Nash equilibriums, to be precise) is also part of our definition of a coordination game, because otherwise the problem of coordination becomes trivial: the players make the only possible choice. If we both need to get across a river, and the only available boat requires both of us to row it, then presumably, if we're rational, we both get in that boat and row. There's no coordination problem to be solved. Whether or not we've entered into a contract to use the same boat is moot. A coordination problem arises only when there's more than one boat, and we must collectively figure out which boat to head for.

In a game of *pure* coordination, the players have all their interests in common. In a game in which the players have more mixed motives, an "impure" coordination game, they have some common interests and some divergent ones. For example, they may be able to procure a prize only if they coordinate their actions, but there may be various ways of dividing it up, or they may be jockeying for position in traffic. If each of us is the same distance from each boat, then picking a boat to converge on is a pure coordination problem. If some boats are more conveniently located for one of us and others are more convenient for the other, then it's an impure coordination problem.

It's here, in dealing with coordination problems, that we encounter a very puzzling and very revealing phenomenon, a phenomenon with much to tell us about human nature at its most distinctive. That phenomenon is the *obvious* and its invisible but tyrannical power over human affairs.

What gives obviousness its strange power? Let's say we can't see which boat the other party is heading for, but we have to meet at the same boat, and one of the seven available boats is bright red, while the other six are all blue. Then the red boat is the most obvious choice, so the easiest way to solve the coordination problem is for both of us to head for it, assuming that the other party will also head for it, because she expects us to head for it, because she expects us to expect her to head for it, because she expects us to expect her to expect us to head for it. This uniquely distinguishing information is so useful that we're both likely to head for the red boat even if it's somewhat farther away from one of us than one of the blue boats and even if that choice is somewhat inconvenient for one party. Clearly, if the problem recurred in this form over and over, it would be easy to establish a tradition of always heading for the red boat.

One of Schelling's simplest examples of a coordination game is "divide the dollar," a game in which two players can share a dollar if they can agree on how to divide it. This isn't a "pure" coordination game, because the players have both common and conflicting interests. Like the problem of converging on the same boat, this is a game of mixed motives, what I'm calling an "impure" coordination game. To get anything at all, the players must coordinate their actions, but within that constraint, they have conflicting interests. Theoretically, one player could hold out for ninety-nine cents, reasoning that the other player would still regard a penny as better than nothing. Schelling observed that human players seldom actually divide the dollar in that way, however. An even division seems to be much easier to agree on, even though it's only one of many possible equilibriums. People arrive at this outcome even if they're not allowed to negotiate or communicate before putting in their bids, presumably through some process of tacit bargaining in which each individual imagines the other individual choosing a percentage, on the basis of the way he imagines that the first individual will choose, on the basis of his expectations about the way the first individual will expect him to choose.

Somehow its very obvious obviousness, its unmistakable salience, the fact that it's the very first solution anyone would think of anyone thinking of, gives an even division an attractive power that's difficult to overcome. Since any division is better for both players than nothing, the most important problem they face is both arriving at the same division, and the lack of a sufficient reason to deviate from the most obvious one makes an even split the choice they both tend to default to. This tiny, apparently ineffectual, inconsequential power, which seems like almost nothing—the power to be the very first idea that happens to occur to people and, perhaps more important, the very first idea anyone would expect anyone to expect to occur to anyone—in fact completely determines the outcome of the game, just as a very similar form of obviousness completely determined which boat we both would head for.

Schelling argued that this attractive power of the obvious would persist even if the players were allowed to bargain explicitly. Each player would expect the other player to expect him to expect her to be unimpressed by arguments that were supposed to show that he should get more than half, and they would still need to converge on one of the many possible outcomes without an inordinate amount of wrangling and confusion, so that's still where they would end up. Human players, even human children, can often converge on this sort of coordination equilibrium without having to have all this explained to them or needing to discuss why it's the logical choice or, if asked, even being able to explain why it is.

Another example of the power of perceived obviousness is the solution that two people might arrive at who are given the assignment of meeting in New York the next day but are forbidden to contact each other in advance to discuss where or when. Many of us would go to Grand Central Terminal at noon and, once there, would look for a conspicuous landmark—the information booth, perhaps—to stand next to. We might well succeed in tacitly coordinating our locations just by picking the most obviously obvious time and place, the place we might reasonably expect the other to expect us to expect him to expect us to be.

To accomplish this, we must do the most obvious thing we can think of. By doing a few simple experiments, Schelling discovered that humans are quite good at this sort of convergence on the obvious coordination equilibrium. They're quite good at anticipating which of a range of possible choices will stick out in a conspicuous way to everyone involved and tend to expect one another to take that one, all other things being nearly equal, because it's the most obvious choice, and coordination on *some* choice is often practically important.

Once our friend knows we'll be looking for him, his own complex and unknown preferences aren't what matters. He might prefer to be sitting down somewhere, eating lunch, but we can reliably expect him to expect us to expect him to stand next to the information booth at noon, even if that's slightly inconvenient for him. If he *didn't* know we were looking for him, or he didn't know that we knew he knew, he wouldn't expect us to expect him to expect us to expect anything, so finding him would become much harder.

In solving these sorts of problems, people depend on the existence of common knowledge. Even though my friend might have expected me to want to go to the Metropolitan Museum, he can't be sure I will have expected him to expect that; he may not be sure that I know that he knows that that's what I'd prefer, or would expect him to respect my preferences rather than his own. But we both know that it's a generally admitted and widely acknowledged fact, known by all to be known by all, and so certainly known to both of us, no matter who we are or what else is going through our minds, that Grand Central Terminal is an extremely obvious and central place, so that's where we each expect the other to expect us to go. If we both want to choose the same point to converge on, without any consultation, it's easiest to rely on such things, on items of knowledge that already have an independent public existence, that already are common knowledge and generally known to be common knowledge, instead of trying to make up new, shared higher-order expectations on the fly.

In this example, one way of solving the problem seems like the most obvious choice, but sometimes none of the boats are red, or one is red and one is yellow. When it isn't completely clear where any rational person would expect any rational person to expect any rational person to expect to meet, or what choice anyone would think of as most obvious, something like a conserved precedent is necessary to resolve the tie. The power of a referee or mediator, Schelling suggested, must often come from the same source. By making one way of settling a dispute the most obvious one, a third party can make it unavoidable by making agreement on anything else impossible.

As the number of players involved in a coordination problem increases, coordination can easily become more difficult, since each player has fewer resources to devote to anticipating the actions of each other player, and that other player's expectations about the other players' expectations about what the other players will do, and more chances to be wrong. The number of possible states of a game like this, a game involving Lewisian higher-order expectations, undergoes a combinatorial explosion as the number of players rises, since every permutation of all the possible expectations, and expectations about expectations, of all the many players might need to be considered. The potential for coordination problems to become complex is one reason that people often need to find some uniquely obvious and unmistakable focal point to converge on. This is one reason it matters what's obvious.

In any m player game, even if there's only one action that the players can take, there are $(m - 1)^n$ beliefs that each player might adopt regarding the nth-order expectations of each other player about who will or will not take that action. In an eleven-player game requiring the attribution of second-order expectations, I must choose from among a hundred possible second-order expectations that I could attribute to each of the other ten players. I may believe that any one of the other players will expect any one of the players aside from herself to expect any one of the players aside from themselves to do something. I might suppose that Joe expects Frank to expect me to do x or that Sally expects Sam to expect Sally to do it. A hundred such sequences are available to me for each player, a hundred for Joe, a hundred for Sam, a hundred for Sally, and so on. Each can be permuted with any of a hundred such sequences for each of the other players. Consequently, my total set of options for simultaneously making such assignments to all ten other players includes 10^{20} distinct possibilities, even without attributing multiple expectations, or multiple possible actions, to single players. Since the universe is less than 10^{18} seconds old, in actual interactions among the members of moderately large human groups, there obviously is no time to consider every possibility.

Some way of drastically simplifying the problem of anticipating what the others will do, and expect us to do, and expect one another to expect one another to do is clearly needed if we're to succeed in coordinating our actions. The power of obviousness (the power of everyone's expectations about what any person whatsoever would expect him to expect anyone else to expect) in determining which possible equilibriums are realized should only grow as the number of individuals trying to solve a coordination problem increases. Without it, the odds of converging on a solution that's optimal in any sense should only decline.

Why must men wear neckties in formal situations? Partly because we all want to do what everyone else is doing, and there are too just many of us to make negotiating a change in the conventions about men's clothing practical at any given moment in time. A conspicuous or authoritative individual—Beau Brummell or the king—might succeed in changing the fashion anyway, however, if he can find a clever way to manipulate what seems obvious to everyone.

Precedent alone is often sufficient to make one of several possible ways of doing things seem more obvious than its rivals, especially when large numbers of people must coordinate on a single choice. To borrow an example from Lewis, once people start driving on the right-hand side of the road, each time you have to decide which side of the road to drive on, your goal being the avoidance of head-on collisions with other cars, the most obvious assumption is that all the other people will assume that all the other people will assume that the safest bet is to assume that everyone else will continue driving on the right, and will themselves continue driving on the right, so you had probably better drive on the right as well. Sometimes when one side of the road is very crowded and the other side is almost empty, it might seem that it would be more efficient to bend this rule a little, but in fact the outcome of such improvisation, if everyone were free to engage in it, could very easily be gridlock or chaos.

Schelling called this sort of arbitrary but obvious clue to successful coordination a "focal point." Conventions, as Lewis ([1969] 2002) conceived of them, involve focal points deriving specifically from historical precedent, which must meet the following five additional conditions:

> A regularity R in the behavior of members of a population P when they are agents in a recurrent situation S is a *convention* if and only if it is true that, and it is common knowledge in P that, in almost any instance of S among members of P,

1. Almost everyone conforms to R;

2. Almost everyone expects almost everyone else to conform to R;

3. Almost everyone has approximately the same preferences regarding all possible combinations of actions;

4. Almost everyone prefers that anyone more conform to R, on condition that almost everyone conform to R;

5. Almost everyone would prefer that anyone more conform to R', on the condition that almost everyone conform to R', where R' is some possible regularity in the behavior of members of P in S, such that almost no one in almost any instance of S among members of P could conform to both R' and R.   (78)

For example, (1) almost everyone in North America drives on the right-hand side of the road; (2) almost everyone expects almost everyone else to drive on the right-hand side of the road; (3) almost everyone cares very little about whether people drive on the left-hand side of the road or the right-hand side as long as almost everyone does the same thing; (4) almost everyone prefers that each other driver adheres to the same rule of driving on the right-hand side of the road as long as everyone else is still doing so; and yet (5) if we all drove on the left-hand side, almost all of us would prefer that each other driver drove on that side too. And all of this is common knowledge, so we can safely assume that most of the other drivers on the road know it, and know that we know it, and know that we know that they know that we know it.

A convention like this is a tool for coordination, a device to allow the awkwardly vast number of possible ways of coordinating our behavior and expectations to be filtered down to a few conventional possibilities. People adopt conventions because they're useful, because without them they would have to think about too many things, because without the convention of driving on the right we would instantly have a traffic jam. This doesn't mean that once accepted, they always stay useful forever. This is the way human social life often seems to be organized: it's a set of incredibly complex coordination problems that we deal with partly by conserving and incrementally improving a set of obvious, though sometimes arbitrary, solutions.

Convention, in this sense of the word, is a pretty good model of the way that a simple signaling system might work in a human population. Suppose there are two possible states of affairs (the English are coming by land or by sea) and two possible signals (hanging one or two lanterns in the belfry) and two possible actions by the receiver (blocking the land route or the sea route). Suppose also that the sender would prefer the receiver to take the appropriate action, and that the receiver would prefer the same thing, so they have common interests. Finally, suppose (contrary to history) that this is a recurrent coordination problem that many different pairs of senders and receivers in a population encounter over and over again.

If there's a famous precedent of hanging one lantern in the belfry if the English are coming by land and two if by sea, this can become a convention if almost everyone goes on doing things that way because the precedent makes it the obvious choice, if almost everyone expects that almost everyone else will go on doing things in that way, if almost nobody cares whether it's one lantern or two that indicates that the English are coming by land, if almost everyone prefers that each other person use one lantern to indicate a land route and two to indicate a sea route as long as most people are still doing things in that way, if almost everyone would prefer that each other person used two lanterns to indicate a land route and one to indicate a sea route if, counter-factually, most people did things in that way, and if all this is common knowledge, so that everyone can safely assume that most other people know it, and know that they know it, and know that they know that most other people know it.

The system can persist through tacit coordination because everyone is trying to anticipate what everyone else will do, and expect *them* to do, and expect them to expect others to do, and will conform without need-ing to have the rules explicitly described to them if they can manage to guess, by observing the behavior of others, what's expected of them in particular situations.

Perhaps some people have only a limited vocabulary. If the British are invading by sea and two lanterns are lit, they'll go out to meet them, but if only one lantern is lit, they'll stay home and do nothing, think-ing that's normal. We would not consider these latter people to be fully competent speakers of the language. For a fully competent speaker to be a true member of the population in which some convention holds, he must, in some sense, know that it holds. A member of that population, the population defined by common knowledge of the convention, "knows" the convention in the sense of being party to it, knowing how to follow it, and expecting others to know how to follow it, and expecting them to expect him to know how to follow it.

Since the purpose of this particular signal is thwarting the British in a way that may be quite risky for those who *do* go out to meet them, and may be made more risky by the failure of those who don't understand the signal to show up, other members of their community may end up being quite annoyed at the people who do nothing whenever they come by land, and may reproach them. The convention itself could always be otherwise, but compliance with the existing arbitrary convention may still be obligatory.

THE DYNAMICS OF CONVENTIONS

If conventions evolve over time, they must come and go. The existence of alternatives—driving on the left or driving on the right—makes this seem somewhat plausible. But what does a Lewisian convention look like as it's coming or going?

Lewis ([1969] 2002:78–80) tells us that there are basically six ways a regularity, R, can cease to be conventional, or not yet quite be a convention, in a group. That we ought to do R—that when a phone call is cut off, the original caller ought to be the one who calls back—might not yet be commonly known by the members of the group of people who are supposed to be party to the convention. If it isn't Lewisian "common knowledge" among them, if it isn't true that almost everyone knows that almost everyone knows that's what you're supposed to do, obviously it can't be one of their conventions. But there are five more ways a practice could fail to be fully conventional. Everyone might know that everyone knew that R was supposed to be the custom and yet

1. Many people might not choose to conform to R anyway—they might just try to call back no matter what because they're impatient.

2. Many people still might not expect almost everyone else to choose to conform to R—they might not know if the other party would really call back when she was supposed to.

3. Many people might not have approximately the same preferences regarding all possible combinations of actions; many people might greatly prefer to conform to R'. For some reason, a large part of the population could greatly prefer that the person who was originally called be the one who was supposed to call back, perhaps because it showed willingness to continue the conversation.

4. Many people, while preferring to conform to R themselves, might also prefer that other individuals not conform to R (not go to Coney Island on Sunday), even if almost everyone else conformed to R, or might be indifferent.

5. Many people might not prefer that other individuals conform to R' (go to Fire Island) even if almost everyone now at Coney Island were to conform to R' (to go to Fire Island instead).

For example, suppose some new notation is introduced for logicians. If nobody knows about it, or if those who know about it don't know that others know about it, or don't know that others know others know about it, if the new notation isn't yet Lewisian "common knowledge," its use can't yet be fully conventional. If everyone knows they're supposed to use it, but nobody thinks it's worth the effort involved so nobody bothers, its use isn't yet conventional. If nobody expects anybody else to use it, for this reason or some other, its use isn't yet conventional. If many people would rather not start using it, its use isn't yet conventional. If most people who use it would prefer that others not be able to use it or understand it, its use isn't yet conventional for the community around the clique of users. If most people prefer that whatever notation they used, this or any other, remained the exclusive property of a clique, then no notation, and in particular not this notation, can yet be fully conventional.

All these obstacles to full conventionality might later be removed. Lewisian "common knowledge" of the new notation might spread; people might begin to use it themselves; they might start to expect others to use it; they might begin to prefer to use it rather than the alternatives;

they might begin to prefer that others use it; and they might become willing to share with the whole community whatever notation or idiom they used. Or a conventional notation might cease to be conventional by losing one or several of these properties, as the logical notation of the *Principia Mathematica* has done in the century that's passed since its introduction.

Individuals can become party to a new convention by meeting the conditions implied by the six criteria, or fall out of the community of full parties to a convention by ceasing to meet one or more of them. They may still continue to follow the precedent out of a sense of moral obligation, or without conscious thought, or in some other way, but once their participation in any of the six ways described—knowing the precedent and knowing that others know it and know that they know it, voluntarily choosing to conform to it, expecting others to choose to conform, preferring to conform if others do, preferring that each other individual conform if everyone else is conforming, and preferring that they conform to some other precedent if that was the convention—ceases to be self-interested and deliberate, they are no longer adhering to the convention *as* a convention.

Each way of becoming, or ceasing to be, a party to a convention requires deliberate human choices or preferences or knowledge, or more or less rational expectations about the choices, preferences, and knowledge of others, even—in the case of logical notation—the choice to become familiar with it, which in this case involves a little labor and, for some people, may require the participation of a teacher. A dynamic version of Lewis's model, in which conventions come and go and therefore evolve, would presumably involve these six processes, either moving forward or going in reverse.

Since all of them involve human knowledge or preferences or rational human choices or expectations, it's hard to see how any evolutionary model other than domestication could be consistent with this theoretical framework. For example, both Russell's ([1921] 2010) behaviorism and Daniel Dennett's (2009a) "synanthropic" model of human language seem to be clearly incompatible with it. We can't both choose to conform to some convention and conform to it inadvertently, having learned it through an unconscious process of association. We also can't adhere to it voluntarily if we have no choice about whether or not to be infested by it. It's we who adopt conventions and we who abandon them, just as it's humans who adopt or abandon dogs.

OUR KNOWLEDGE OF CONVENTIONS

Nevertheless, we shouldn't overestimate the amount or kind of knowledge necessary to be a party to a convention. Even people who are full parties to the conventions of a particular language or dialect, who are able to use and respond correctly to all the community's conventional signals, may still have many different degrees and kinds of knowledge. To be party to a convention, we must "know" it in the purely practical sense of being able to conform to it if we so choose, but Lewis ([1969] 2002) warns us that this implies that our knowledge of conventions may be "quite a poor sort of knowledge" (60–68).

We can "know" a convention, he argues, if we're merely in a position to believe that it holds, should the question ever come up, but haven't yet formed any such belief. We can know conventions in irremediably nonverbal ways or on the basis of evidential justifications that we could never describe or report; and we may know how to follow a convention on the basis of our knowledge of the many particular things or situations that fall under it, without being able to assemble that knowledge into the sort of general claims about categories or kinds of thing that would allow us to describe the convention we're following in general terms. Borrowing terms from Abelard, Lewis tells us that we may know the convention in *sensu diviso* rather than in *sensu composito*.

We may know in some inchoate way that all the adults in our small town habitually drive on the right side of the road that goes by our home, without knowing that all drivers on all roads in our country must drive on the right. Knowledge of the features or behaviors of particular drivers, however many, doesn't directly support inductive inferences about which features drivers in general necessarily must have. There may be a way to get from the first kind of knowledge to the second, but some additional hypothesis about the existence of generalizable classifications such as "driver" and "right side of the road" is required in addition to a list of the proper names of particular persons and locations with their many individual features, and this is precisely what the knower of a general convention in *sensu diviso* lacks. He's capable of recognizing all the cases to which the convention applies, or all the ones he's likely to encounter, and he knows what everyone knows one ought to do in each case, but he can't explain which common features make the cases the same. He doesn't quite know which criteria of identity (in the sense of the term explained in Slote [1966], see Lewis [(1969) 2002:24n.4]) are the practically important ones.

The kind of *sensu composito* knowledge called for in this example is so trivial that only a child could lack it, but we should remember Justice Potter Stewart's definition of pornography: I can't intelligibly describe it, but "I know it when I see it." Many of us would say more or less the same thing about the ethical way for a member of our community to behave, or what we should accept as fine art, so not all our *sensu diviso* knowledge concerns trivialities, or things whose correct description is a mystery only to children. "What, then, is time?" Augustine asks in the *Confessions*. "If no one asks me, I know; if I want to explain it to someone, I don't know." Yet we all manage to use the word, so we must know its meaning in *sensu diviso*.

It might seem that this is stretching the meaning of the word *knowledge*. We're used to thinking that anything we know, we can declare—but do we really want to take the position that a person *doesn't* know what the word *time* or *because* means, even if he can conform perfectly to the conventions governing its use, unless he can also state them explicitly? Do we really want to say that nobody knows what the word *know* means unless she can give a full and explicit account of its meaning? The claim itself seems to be self-undermining: wouldn't we have to know what *know* means in order to conclude that we don't "know" what it means?

In that case, almost nobody knows those words, so it's odd that we all can use them and spot mistakes in others' efforts to use them correctly. ("You can't say you're late 'because' it rained, you started late so you would have been late anyway." "You don't *know* that, you haven't even looked yet.")

It seemed better to Lewis to say that we know their meanings in a certain inchoate way, know them but only in *sensu diviso*, since this is the only way past Quine's supposition that a convention can be adopted only if those choosing to follow it can state it explicitly and all the unpalatable consequences that seem to flow from that assumption. It seems better simply to accept that rational choice doesn't necessarily presume declarative knowledge. In fact, things seem to be the other way around, at least for language. We know the meanings of words. We can choose between the word *know* and words like *believe* or *hope* or *suspect* in a more or less rational manner. First, there must be knowledge and a word for it, and words for its alternatives, and some inarticulate sense of the nature of this thing, "knowledge." Producing a final, explicit, accurate full description of the meaning of the word *knowledge* is a rather late step in the process of coming to know what that word means, one we haven't actually reached yet in this particular case.

It isn't clear why this should surprise us. First there had to be gold, and a word for it, and words for brass and bronze, and some knowledge about this thing, "gold." Our current explicit definition of *gold* as "the element with atomic number 79" could have come along only after many centuries of successfully using the word to refer to gold. Still, Pharaoh knew what his language's word for "gold" meant, as anyone who tried to sell him brass while calling it gold would soon have discovered.

THE LANGUAGE OF A HUMAN POPULATION

However we know it, a set of conventions about hanging lanterns in windows to indicate which route the British are taking is a signaling system. A verbal signaling system could also be constructed along the same lines. According to Lewis ([1969] 2002), such a system is a language with the signals as its sentences, but it's a rudimentary language, much simpler than the language of any actual human population. A fixed, finite set of signals can be sent to identify a fixed, finite set of situations. Each situation should be responded to in one particular way, or at least with one particular discretionary contingency plan, and both parties want that equally. There are no questions, there's no conversation, and nothing new can ever be said. There's no syntax.

At this point, what the verbal signaling system already *can* have is an interpretation. For each signal of the system, this consists of a pair $\langle \mu, \tau \rangle$ of a mood and a set of truth conditions.

As examples, Lewis picks two moods, the declarative and the imperative. In this rudimentary language, signals activate contingency plans on the part of the hearer, and result from contingency plans themselves: If you see an *s*, say "σ" or if you see an *s*, say either "σ" or "v," or something like that. The difference between the declarative mood and the imperative, he argues, depends on the relative amount of latitude in the sender's and receiver's contingency plans. On observing the British coming by land, if the sender has no option but to put one lantern in the belfry, but the Minutemen, once warned by this signal, have a variety of responses they might take, the signal is a declarative one. But if the sender has several choices about the number of lanterns that he might hang, depending on what his overview of the situation suggests is best, and the Minutemen are restricted to responding to each of the possible signals in a particular way, then the signal is imperative.

The truth conditions of a declarative sentence are just the set of ways the world could be that would make it true, the set of "possible worlds" in which it's true. To determine whether it's true of the actual world, we construe the words of the sentence as referring to things, or sets of things, in the world around us. We have to fix their reference by correctly relating kind-names to instances of that kind of thing in the world, interpreting indexical terms like *that*, *here*, and *my* as referring to particular individuals, things, or places and so on. The speaker's declaration "My gold earrings are in the upper-right-hand drawer" must be related to a particular set of small metal objects boxed up inside some particular planes of wood. Even a sentence like "If Nixon had never been president, he never would have been impeached," which is about events in a merely possible world, still requires us to look through the history of our own world for the referent of the name Nixon and then to imagine that world changing around him. Once we've interpreted the sentence as a claim about particular objects, events, and so on and determined whether it's true in the real world, we can start to figure out what other sorts of worlds it could be true in, keeping the earrings and the drawer, or Richard Nixon, fixed in our mind and allowing the world to change around them. So as I am using the word here and in subsequent chapters, the verb or action, *interpretation*, is the process of arriving at *an interpretation*, the thing, $<\mu, \tau>$. It is the process of finding the right things in the right worlds, and figuring out what other worlds the sentence would still be true in.

The consequence of a true declarative sentence being uttered, heard, and believed is that our shared conception of the world we live in (and/or what other possible worlds are like) is altered. We come to believe that the specific pieces of yellow metal being referred to, the ones we saw yesterday, are indeed boxed up inside the glued-together pieces of wood. What to do about that is up to us. It's the sender's responsibility to make sure a declarative sentence is true, to send it only when the actual world, or the possible world under discussion, is one of those in which it would be true.

For Lewis, the truth conditions of an imperative sentence are also the set of worlds in which it's true. Now, however, it's the receiver's responsibility to make sure the sentence is true, to make the actual world into one of those worlds. When I say "Scalpel!" in the imperative mood, I expect to soon inhabit a world that contains a scalpel in close proximity to myself, and will be disappointed if I don't eventually see one. I'm not modifying our shared picture of the world by uttering this instruction; instead, I'm asking *you* to modify the world itself.

What gives the human speaker an incentive to make declarations only when they're true of the actual world, and what gives the human receiver an incentive to make the actual world into the one described by the speaker in his instructions, is the existence of common interests, which give the speaker the authority to direct or to declare, and the receiver an incentive to obey or to believe. In the absence of a common interest in the instruction being obeyed, Lewis feels that orders, since they presumably must be enforced with threats, are better interpreted as promises, a type of declaration. ("Get me a brick, or I promise you'll be sorry.")

Since the concept of "common interests" lies at the roots of our ability to communicate in this very constitutive way, it would seem to be fundamental to our whole way of thinking about human uniqueness. A sense of having common interests in the performance of shared tasks is precisely what experts on their behavior like Michael Tomasello (2008:38–55, 172–85) tell us that chimpanzees most strikingly lack. Lewis's way of conceiving of the common interests involved in adhering to a convention, however, is more complicated than it might appear.

How similar must our interests be? It may be good for you if the order is obeyed or if we both adhere to the convention, and even better, much better, for me. Even though you may be threatening me with retaliation if I disobey, both of us might prefer that you didn't have to carry out your threat, so even credible promises of harm can *create* common interests. Most people may follow a convention because they want to, while a few do so only because they fear punishment for unilaterally deviating. Do all these people have a common interest in following the convention? To what extent must interests be identical, or similar, for a signaling system of this Lewisian kind, or any other set of Lewisian social conventions, to exist? If some conventions are protected by the threat of punishment, are they still merely conventional? For example, isn't the speaker's obligation to say things that are true in his language part of the social contract, rather than part of the conventions of the language? Don't we sometimes punish fraud and perjury? In doing so, are we punishing the mere violation of an arbitrary convention? Does a liar merely violate an arbitrary convention?

Lewis had an answer for these questions. He explained that the interests of the parties to a convention do not need to be exactly identical, as they would have to be in a pure coordination game, but only "nearly identical," in the sense that no one individual could do better for herself by not participating at all, by simply not coordinating her behavior with that of the others, by just refusing to play by the rules ([1969] 2002:13–15, 88–97). Speaking Italian must at least be better for each person who wants to participate in the conversation, given that everyone else is speaking it, than not participating at all, in order for it to be a convention among them that the conversation will be conducted in Italian.

Conceivably there are subgroups of *more* than one individual—up to and including the whole group—that would be better off if *none* of them adhered to the convention. If all the workers in an unsafe mine stayed home, they all might be better off, being in a better position to negotiate for safer conditions, but nobody wants to stay home all by himself and end up being fired while the others work. And as long as we, and most of the others, are still going to work, we want each other worker to be there as well, in order to make the job less dangerous. So some situations that are very bad for some or all of us can still be situations in which we'd all individually choose to stick to the governing convention. As Lewis ([1969] 2002:92) says, sometimes we're all trapped by a convention. Once created, conventions may be very hard to escape and very hard to uproot.

It's in these sorts of cases that a convention can be inconsistent with the social contract. The social contract itself, according to Lewis, differs from a convention in that we all would prefer that *everyone* adhere to it, that everyone refrain from murder, whereas a convention requires us only to prefer compliance to unilateral nonconformity by ourselves, or by any single other individual, given that everyone else is conforming. We prefer going to work in the mine to being the odd man out, and if we all are working, we won't want other individuals to be idle, but it would be much better for everyone if nobody went to work, if we all went on strike.

This makes the regularity of continuing to work in the mine very different from the regularity of telling the truth in a human language, since it would be difficult to make the case that we all would be better off if we all lied all the time, or all told the truth in our own distinctive brand of Italian, in which the meanings of the words were gerrymandered in whatever way happened to be most beneficial to ourselves. In fact, almost all of us would prefer that everyone told the truth in a fairly standard version of the language of our community, a preference that seems to give conventional human languages a special relationship to the social contract. The social contract doesn't mandate that we speak Italian rather than Welsh, but it does have a special relationship to the general understanding that when we represent ourselves as speaking Italian, we should speak the same version of that language as everyone else does, and tell the truth in that version, instead of lying.

As the example of driving on a public road suggests, many conventions are associated with clubs, which use penalties or exclusions to prevent nonconformists from participating in the activity governed by the convention, or to force them to conform if they do participate. But the convention of driving on the right still isn't the social contract. If someone started a political movement that advocated changing to driving on the left, it wouldn't be treasonous, because there would be no attack on peaceable and orderly coordination among the members of society in general. If the movement succeeded and we made the change, some people would be inconvenienced, but life would go on. In contrast, unilateral noncompliance, driving down the wrong side of the road at high speed, is likely to result in negligent homicide, so it *is* inconsistent with the social contract.

The real difference between a set of conventions and a social contract, Lewis argues, is that the set of conventions has this sort of alternative—driving on the left rather than the right, or speaking Welsh rather than Italian—in which players are still coordinating, an alternative that is, in his sense, "almost equally good" for almost everyone. There is an alternative that is still better than unilateral violation, still better for almost everyone involved than being the odd man out, even though for some members of the community, it may in fact be considerably better or worse in an absolute sense than the convention we're actually following. In contrast, the social contract has no alternative except a Hobbesian hell. We prefer that people drive on the right because it makes coordination possible in a complex world, so the preference is conditional on everyone else doing the same thing. But we care about the prohibition against murder for its own sake, no matter how many people happen to be doing it.

The rules of the club of drivers derive their authority from the social contract, not because they're among its actual clauses, but because there must be *some* set of rules if the club is to function in a way *consistent* with the maintenance of the social contract. (If the functioning of the club is inconsistent with the maintenance of the social contract, if it's a gang of criminals, that contract might actually forbid us to follow its rules.) We accepted our driver's license with the understanding, between ourselves and the other drivers, that we would obey the rules of the road, whatever they might be. The convention of driving on the right is just a convention, but it is nevertheless a binding one. There are always conventional annexes to the basic social contract.

Since most of us do want to continue using whatever language our community uses, and since the use of some language or other is indispensable to maintaining the social contract, we must remember that we've been admitted to the club of informative speakers of that language with the contractual understanding between ourselves and the other speakers that we would adhere to its conventions, not to other made-up ones, and are not supposed to surreptitiously or flagrantly violate them by, say, using words in very nonstandard and self-serving ways or quibbling invidiously after the fact about the meaning of some word we used, or by uttering falsehoods or culpably misinterpreting what's said to us, without warning anyone that this is what we're doing (Lewis [1969] 2002:97–100, 177–95): "A convention of truthfulness in L is a social contract as well as a convention. Not only does each prefer truthfulness in L by all to truthfulness in L by all but himself. Still more does each prefer uniform truthfulness in L to Babel, the state of nature" (182). Lying in English is a violation of an obligation we have undertaken voluntarily by choosing to speak English. Willful or negligent misinterpretation is a violation of obligations we've voluntarily undertaken by presenting ourselves as competent members of an English-speaking audience. Telling the truth and faithfully interpreting what's said to us are part of the rules of the road.

This relationship to the social contract, and the mechanisms of enforcement associated with many conventions, suggest that thinking of the Lewisian conventions of a human language as simply being equilibriums in a coordination game isn't entirely correct. Cooperative games are precisely those games in which players can make binding contracts before the start of play. The social contract is a binding one. So this relationship with the social contract turns Lewis's "coordination games," at least insofar as they're supposed to be models of the conventions of language, into cooperative games, a rather different thing. Not only does each person prefer each other person to use the word *gold* only for gold, provided that everyone else is doing so, but almost everyone prefers that everyone use the word to mean the same thing. The conventions aren't just convenient precedents established in the interactions of particular pairs of people; they're also norms backed up by the approval of the community as a whole.

We will see the alternative—precedents established in the interactions of particular pairs of individuals, without any role for the community as a whole, signaling "conventions" not associated with a social contract—when we look at the gestures that chimpanzees use to communicate (Call and Tomasello 2007).

In the picture of human behavior that Lewis gives us in *Convention*, most people follow the reigning conventions of the community's language out of self-interest most of the time, without being forced to do so by the threat of punishment. Those people still do use some version of the sort of reasoning Schelling suggested that we use to solve the *informational* problem of coordination, to discover what coordinating would consist of. In some sense, then, they're playing a coordination game *inside* a cooperative game. But those who blunder or choose to experiment too freely will find that the Lewisian conventions of a human language are actually binding ones. (Putnam made the same general point very forcefully in "The Meaning of 'Meaning'" [1975b:248–49.]) When it comes to things like calling gilded bars of lead "gold," the enforcement of these binding conventions may become quite draconian.

The conventions that organize a human society often are binding for a good reason. In the decades after Lewis wrote *Convention*, Michihiro Kandori, George Mailath, and Rafael Rob (1993) and H. Peyton Young (1993) demonstrated that that simply letting nature take its course in a coordination game like this, a game involving temptations to misbehave, tends to make you end up at a "risk-dominant" equilibrium that can be far from optimal and may not involve any coordination at all.

Rousseau's famous story, in the *Discourse on the Origins of Inequality* (1754), about a stag hunt is an example that's often used to illustrate this problem. (Lewis [(1969) 2002:7] uses it as an example of a coordination problem in his introduction of the subject.) A group of hunters can capture a stag if they all work together. This is the best outcome for everyone. But sometimes while they're hunting, a rabbit runs past. If one hunter deserts his post to chase the rabbit, he's quite likely to catch it, which will give him some meat to eat, even if not as much as he would have gotten from the stag. It's best for everyone if he ignores the rabbit, since that will lead to the highest possible payoff, provided that everyone else also does their part—but what if one of the other hunters decides to chase it? Then he'll get nothing.

If none of the hunters ever makes the mistake of chasing the rabbit, if everyone is always completely certain about the relative payoffs of the possible outcomes to everyone involved and the complete rationality of all the other hunters, then the coordination equilibrium of hunting the stag will persist. But suppose that one hunter tries a foolish experiment, or mistakenly believes that he can get more by chasing the rabbit, or irrationally decides that the other hunters are unreliable and goes after it. Then the other hunters' confidence will be damaged. It will no longer be rational for them to be certain that everyone else will do his part, which means that it's now risky for them to do their part themselves, meaning that it's risky for them to expect others to do their part, or to expect them to do their part, or to expect them to expect the others to do their part. Recursive mind reading of the kind that Lewis describes can help them solve the informational problem of how to succeed in coordinating if everyone is actually trying to coordinate, but it can't predict blunders or unwise experiments, so it isn't clear how it could solve this problem.

In this example, both parties chasing the rabbit is what John Harsanyi and Reinhard Selten (1988) dubbed the "risk-dominant" equilibrium, even though both parties chasing the stag is the "payoff-dominant" equilibrium, given some reasonable interpretation of the problem in terms of utilities. Kandori, Mailath, and Rob (1993) showed, well after Lewis wrote *Convention*, that evolutionary versions of this sort of game allowing occasional mistakes or experiments would, if the likelihood of further experiments eventually declined to zero, almost certainly end up at the risk-dominant equilibrium of noncoordination, and not the optimal payoff-dominant equilibrium of everyone chasing the stag. (If the players continued endlessly experimenting, they wouldn't "end up" anywhere, which is almost as bad from the point of view of sustaining their interest in trying to participate in the stag hunt.) In "The Evolution of Conventions" (1993), published in the same year, Young demonstrated a similar result.

For Lewisian conventions, the solution to this problem seems to be contained in clauses 4 and 5 of his definition. Not only do we *prefer* that each other individual also follow the convention, provided everyone else is; we often take actions, extreme or mild, to ensure this. In the real world, people who can't demonstrate that they know which side of the road to drive on, or simply aren't willing to drive on the same side as the other drivers, are kept off the highway. They aren't allowed to join the club of drivers. Experimentation (if you're in a hurry, driving on the wrong side of the road to see if that would improve your speed) is strongly discouraged. Even innocent mistakes (driving on the wrong side of the road by accident while drunk) are quite likely to be punished. So this particular convention also has contractual force, though it's seldom necessary to actually exclude anyone from the club of licensed drivers for those particular reasons. Kandori, Mailath, Rob, and Young assume that there are no traffic police actively discouraging experimentation with noncooperative behavior and careless mistakes, which makes it somewhat unsurprising that their models predict gridlock. For their stability and continued existence, the Lewisian conventions of an actual human language depend on the existence of a social contract and the mechanisms for enforcing that contract that come along with it.

Most people voluntarily follow the convention of driving on the right *as* a convention rather than as an obligation. It's just that there's also a filter in place to keep out those few who won't or can't, to prevent selfish behavior, ignorance, myopic rationality, and other kinds of noise from disrupting our ability to voluntarily coordinate around the precedent. We might naively hope that if the traffic police permanently vanished tomorrow, people would continue indefinitely to obey the rules of the road. In fact, however, probably a few people would start playing fast and loose almost immediately, for what seemed to them to be perfectly sufficient reasons. Given the large number of people on the road and the vast number of permutations of higher-order expectations that would then have to be considered in real time to avoid accidents, the confusion they generated would eventually undermine the workability of the whole existing system of conventions.

A convention of this kind, which has a filter around it to prevent scofflaws from ruining the coordination that it permits for those who would like to comply with it, must be clearly distinguished from a convention to which almost everyone conforms despite being completely free not to, because nobody ever has a significant incentive to deviate, and nobody ever makes a mistake or tries an experiment. Are there any such nonbinding conventions? Young gives the example of going to lunch at noon, which certainly is nonbinding. But many people *don't* go to lunch at noon—if they're busy, they may not go until 1:30, or at all. Sparse and inconsistent observance of this kind may be characteristic of nonbinding conventions in general, since people are free to ignore them if they wish.

In a static model of the conventions of a human language, the distinction between binding and nonbinding conventions (telling the truth in standard English or driving on the right, versus going to lunch at noon) might not seem to matter enormously, since we may be interested mostly in the typical experience of the typical individual, the individual who conforms voluntarily. In an evolutionary model, however, it seems likely to be very important, since it's precisely at these filters that what does and doesn't count as adequate adherence to the convention is settled, and it's precisely through their action that the distinction might be maintained. By preventing dangerous forms of experimentation and discouraging casual blunders, they can allow the population to stay at the sort of non-risk-dominant optimal coordination equilibrium that Young showed might be stable if, and only if, most such things are absent.

From now on, to make this difference clear, I'll continue to speak of conventions of the first kind, conventions that come with highway patrol officers or disapproving peers attached, to filter out scofflaws or those who might be tempted to call gilded lead "gold," as *binding* conventions. This will allow me to distinguish them from the nonbinding conventions that Young wrote about, which leave us free to experiment or slip up, without any special sanction being applied to those who do.

A Youngian nonbinding convention differs from a Lewisian binding convention in that it doesn't seem to have any associated preference corresponding to clauses 4 and 5 of Lewis's definition. Each agent simply seeks what's best for himself, given whatever expectations about the behavior of others his experience has allowed him to form. If he does have preferences about the other players' behavior, he's in no position to do anything about them. A binding convention is simply a convention for which Lewis's clauses 4 and 5 have real teeth. Of course, when the temptation to misbehave is very weak or the consequences of noncompliance are trivial, the sanctions needed may be trivial as well. They may consist only of displeased or amused facial expressions, sarcastic remarks, or awkward silences.

The convention that we tell the truth in the specific version of our language that we share with most other speakers is a binding convention, and the pathological liar or habitual misinterpreter may soon find that the people he lies to or culpably misinterprets may be inclined to exclude him, if they can, from the club of informative speakers or competent audience members, or to apply some other sanction. It's this quasi-contractual obligation to tell the truth in some fairly standard version of our language that makes its use a cooperative game and not a true coordination game, even though most players (those who seldom lie and are seldom lied to in consequential ways) may have experiences that make it look like a mere coordination game most of the time.

So is any particular language, L, ever actually the language of any human population? According to Lewis, no. In fact, in his opinion we never quite manage to converge on a single language, even though we're always trying:

> I think we should conclude that a convention of truthfulness in a single possible language is a limiting case—never reached—of something else: a convention of truthfulness in whichever language we choose of a tight cluster of very similar possible languages. The languages of the cluster have exactly the same sentences and give them corresponding sets of interpretations; but sometimes there are slight differences in corresponding truth conditions. These differences rarely affect worlds close enough to the actual world to be compatible with most of our ordinary beliefs. But as we go to more and more bizarre possible worlds, more and more of our sentences come out true in some languages of our cluster and false in others.
>   . . . by not committing ourselves to a single language, we avoid the risk of committing ourselves to a single language that will turn out to be inconvenient in the light of new discoveries and theories; we allow ourselves some flexibility without change of convention.   ([1969] 2002:202)

Everyone speaks a slightly different idiolect, even though we are constantly trying to harmonize them all. The result is that our language has the internal resources to change in the face of new circumstances. This, of course, is very Darwinian. Over time, this naturally occurring variation presumably would allow the conventions of a human language to undergo a long series of small, sequential, perhaps almost unnoticeable changes as a result of human experiences and preferences, with one incremental variant repeatedly replacing another slightly different one in a subtler, less explicit version of the same sort of process that Lewis described for logical notation. The endless displacement of existing versions by incrementally improved ones is the essence of Darwinian evolution, so it seems to me that an implicit theory of the evolution of the conventions of a human language is already concealed in the static account of their character that Lewis has given us.

Of course, if we use a word in some slightly nonstandard sense without warning anyone or being willing to admit it, we're likely to run up against the binding nature of the language's conventions. Often, however, we're able to explain what we mean or obliquely indicate that in a way that gives fair warning that we're using a particular variant of its meaning. Still, some parts of this evolutionary process must often involve acrimony, accusations of bad faith, and attempts to pin down precisely what people are saying. The point is just that we don't always end up pinning them down to the exact same place that we ourselves were in before the conversation started. This sort of friction is the inevitable result of the persistence of slightly different idiolects in a system of what are supposed to be binding conventions. It's one of those strange, counterintuitive Darwinian processes that require errors to be continually introduced for perfection to be achieved (Cloud 2011).

There are further complexities to a real human language. An unrealistic aspect of the story as I've told it so far is its restriction to simple signaling languages like "one if by land, two if by sea." These are still much simpler than any actual natural language. They can map only a certain finite set of situations to a certain finite set of contingency plans. Actual human languages, in contrast, have a potentially infinite number of sentences. Each grammatical sentence that isn't nonsensical has a mood and a set of truth conditions, and if the sentence is ambiguous, it may have more than one of each.

Generating an effectively infinite number of sentences is no problem. They can be generated by the right kind of computer. We know exactly what sort of computer would be required to create the sort of finite but unbounded collection of strings of symbols that the actual sentences of human languages fall into. There's no great puzzle about how it's physiologically possible to produce enough distinct sentences, though it is a bit puzzling that no other living thing seems to string sounds together with the same degree of complexity. If it's physiologically possible, why isn't it more common? That, however, is a question for subsequent chapters. Right now, what probably ought to be puzzling us is how we manage to assign each of a potentially unlimited number of novel sentences its own peculiar set of truth conditions.

CONSPICUOUS ANALOGIES
AND NONNATURAL MEANING

What sort of thing is the phrase "the British are coming by sea" supposed to stand for? We never face exactly the same situation twice, Lewis ([1969] 2002) tells us, so all the situations in which the signal of two lanterns is the appropriate one to send aren't exactly the same, and the coordination problem we face isn't exactly the same coordination problem each time. We really should think of the signal as one we should send in any of a large number of analogous situations, situations in which an analogous coordination problem arises and can be solved in an analogous way. The skill of interpretation requires mastering these analogies and learning how to apply them to particular situations. Lewis argued that a language is a system (a weaving together, in Plato's very descriptive metaphor) of many different analogies, and a grammar for stringing them into collages of analogies, sentences that are true or false in possible worlds or sets of possible worlds, sometimes including the actual one, and that we interpret as referring to particular objects in this world or those other worlds.

Of course, any actual situation is analogous to many other situations in myriad different ways. Fortunately for us, Lewis ([1969] 2002:37–38) explains, most of the analogies strike us as artificial, and only a few leap out as "natural." We ignore the artificial-seeming ones, and we expect others to ignore them, and to expect us to ignore them, and to expect us to expect them to ignore them. Because everyone expects everyone to expect everyone to ignore all but the most natural-seeming, obvious, and readily apparent analogies, we all can converge on the ones that anyone would expect anyone to expect anyone to notice.

The fact that already, several times, Redcoats have come in ships to attack our town isn't just a noticeable similarity between different events; it's a similarity we'd expect anyone living in the town to expect anyone living in the town to have noticed. Even if it also was cloudy on every occasion, the unusual but similar occurrence of soldiers arriving by sea each time two lanterns are lit has an obvious salience that trumps that similarity, so the signal is likely to be interpreted as referring to the soldiers, not the clouds, even by those not in on the secret. Once the British commander has been thwarted in this same way several times, he is unlikely to conclude that two lanterns are the local signal for clouds, and one for a clear sky.

This preeminent salience of certain analogies might be of two kinds. Some features of an environment—for example, whether or not any part of it is on fire, in a way analogous to what was happening when we once were burned, or contains food, as it does whenever we get to eat something—would stand out as especially salient to any creature whatsoever. Some things have salience simply as natural incentives. But we humans also might pay more attention to certain features of our environment because we expect others to do the same, and we may expect them to do the same because we expect them to expect us to do the same. Of all the analogies between all the various bits of paper we deal with, the analogies between certain pieces that make them all postage stamps or money are particularly noticeable to us, partly because we expect others to notice them and to expect us to expect them to notice them. Those analogies are salient because their existence and importance are Lewisian common knowledge, known by all to be known by all. We might say that these analogies are made salient by their commonly known "entrenchment" (Goodman [1955] 1983) in the minds of the population of speakers, rather than by being associated with direct natural incentives. In fact, we're required to recognize these particular analogies and not others, to speak only of actual gold as gold and not to use the term, for example, in promises, to refer to our own, very different homemade version.

This is another area where developments in game theory that came after Lewis wrote *Convention* might give us reason to worry that he may have been too optimistic about our ability to spontaneously converge on the optimal form of coordination in the absence of any policing. One problem is that successfully attending to the same analogies and the same environmental cues as everyone else involves work, which may have a cost. Ken Binmore and Larry Samuelson (2006) demonstrated that if nature is simply allowed to take its course—if paying attention to environmental cues is costly, if how much attention the players of a coordination game pay to which sorts of environmental cues is left up to the individual players, and if no sanctions are applied to players who don't pay enough attention or who attend to the wrong thing—then the players won't pay the optimal amount of attention to the right cues. Instead of arriving at the optimal, "payoff-dominant" level of attentiveness to possible cues for coordination, players will gravitate toward paying a smaller, "risk-dominant" amount of attention, even though they may often miss cues that would allow them to coordinate successfully in particular situations.

Basically, the reason is that when a player who is paying close attention to lots of cues or distinguishing marks is paired with a player who is paying slightly less attention, it's the attentiveness of the inattentive player that determines the probability that coordination will be successfully achieved. The inattentive player at least is spared the costs associated with attending carefully, but the attentive player gains the exact same benefit from their interaction, while paying a higher cost for monitoring the environment.

Since slightly less attentive players do better against slightly more attentive ones than the more attentive ones do against them, they can afford to do slightly worse in their interactions with their own kind than more attentive players do in their interactions with *their* own kind. So a population of more attentive players can be invaded by slightly less attentive ones. Through a long series of small steps of this kind, the population will inevitably move away from the optimal, "payoff-dominant" equilibrium of paying close attention to a large set of cues or differentia (which might allow all of us to see a fairly complex set of analogies as conspicuously salient) toward a more inattentive, "risk-dominant" equilibrium in which everyone is worse off, because fewer analogies can be treated as conspicuously salient.

Here again, by analogy with the public highway, the problem apparently would be solved by the existence of some sort of "attention police" who punished people for being inattentive or failing to attend to the right cues, weeding out slackers so that those who wanted to coordinate efficiently would be in a position to do so without interference from inattentive invaders.

Human parents, human teachers, and other humans with whom we interact do sometimes sanction us if we fail to pay attention to the sorts of things we're supposed to be paying attention to. Not paying adequate attention is often grounds for reproach. Certain kinds of conversation—

for example, almost all the conversations that take place at universities—might be less efficient if people who were not very interested in their subjects, and the fine distinctions they require, were completely free to participate in them, without even being frowned at. It seems plausible that the sorts of mild (or severe) sanctions or exclusions that humans sometimes apply to other humans when they fail to pay enough attention to the right things may help us share a richer sense of which analogies are the conspicuously salient ones in particular situations. This may be one reason that teachers give tests or glare at students who make comments in class when they haven't done the reading.

An obligation to pay attention to certain things can be part of the social contract in the same way that the obligation to use the word *gold* only for gold is. Using the word *gold* only for gold is actually rather difficult, since it requires us to pay attention to certain abstruse distinguishing marks, which allow us to tell the difference between gold and fool's gold, so the two obligations are intimately linked.

Some analogies are naturally conspicuous, while others stand out because others expect us to expect them to expect us to treat them as conspicuous, and may be irritated if we show no signs of any such expectation, or don't seem to care what they expect us to do. In a similar way, some signals are naturally informative, and others are informative because others expect us to expect them to expect us to treat them as informative. For Lewis, even signals in a mere signaling language like "one if by land, two if by sea" have the second kind of meaning, what H. P. Grice (1957) called "nonnatural meaning" (or, for short, "meaning$_{NN}$"). The intended contrast—which has absolutely nothing to do with not being part of nature—is with what Grice called "natural meaning."

Smoke naturally means a fire, and spots naturally mean measles, but to mean something *by* smoke is to use a smoke signal. Glancing at my watch may be a sign that I'm out of time, but only if I expect you to recognize that by glancing at my watch I mean to inform you that I am almost out of time do I mean$_{NN}$, by glancing at my watch, that I am almost out of time. A sigh may mean boredom, but meaning$_{NN}$ boredom by means of a sigh involves an intention that the theatrical sigh be recognized by the audience as intended to convey boredom, and the recognition of the intention must be what causes the belief. A nonnatural signal doesn't merely cause a belief about the world; it causes a belief about the utterer's intent to cause a certain belief about the world. An accurate fourth-order picture of mental states on the part of the receiver (he intends me to believe that he intends me to believe) is required in order for Gricean nonnatural meaning to be successfully conveyed.

If I show Mr. X a photograph of Mr. Y showing undue familiarity to Mrs. X, he may conclude that something improper may have occurred, whether or not he attributes to me any intention to communicate that. But if I draw him a picture of the same thing, he will begin to suspect Mr. Y only if he realizes that I'm deliberately trying to send him a message, since the drawing itself is hardly evidence of impropriety. I can stop adding details to the drawing once he realizes what I'm trying to do. The certificate of intent to draw a picture can be much simpler than the whole picture would be: he may get the idea right away and punch me in the nose before the insinuating picture becomes very detailed.

If a policeman stops a car by standing in its path, the driver may slow down just to avoid an accident, but if a policeman tries to stop a car by waving it down, the driver must recognize his intention to communicate his wish that the driver should stop to be able to follow the instruction. Otherwise, it means nothing. Perhaps he's saying hello? Only if we recognize that the policeman intends us to recognize that he wants us to stop, expects us to expect him to expect us to stop, will we stop.

We'd better recognize that, though. The recursive mind reading he's asking us to engage in isn't optional. The policeman will be upset with us if we're distracted by some other feature of the environment and pay no attention to his waving. Failure to grasp his intention that we should grasp his intention may even be a crime, and the next thing we hear may be a siren. Which we *really* had better understand as intended to convey to us an intention to inform us that we must stop, or even worse consequences—a car chase, our arrest—will probably ensue.

Paying attention to the things we're expected to pay attention to is often mandatory. Conventions about which analogies should be regarded as conspicuously salient often are binding. Faithful interpretation is part of the social contract, just as not lying is. We can't just pretend we didn't understand the siren or understood it in our own unique way. Saying that we didn't notice it because our favorite song was on the radio or because we were busy reading a text message won't excuse us.

By driving on a public road, we've accepted the obligation of listening for sirens and interpreting them in a certain way, so to silently renege on that obligation later is to act in bad faith.

The sender of a Lewisian signal intends that the audience's recognition of his intention to communicate something to that audience should be effective in communicating it. Once the intention itself is recognized, nothing else has to be communicated, so the signal can be quite stylized. It can be ringing a bell, or hanging two lanterns in a window, or drawing a very crude and simple picture, or waving a hand in a certain special way, or briefly turning on a siren. But the sender doesn't regard it as a foregone conclusion that even if the intention isn't recognized, the message will still be understood. Only if the receiver understands that by ringing the bell or hanging the lantern or drawing the picture or waving the hand or switching on the siren, the sender is trying to tell her, the receiver, something will the receiver be able to figure out what the sender is trying to say.

Since the attribution of particular intentions to particular other people is an all-things-considered judgment, the receiver's whole picture of the world, and the nearby possible worlds, and of the sender's picture of the actual world and the possible worlds closely associated with it, and of the sender's picture of the receiver's picture of the world all come into play in interpreting a signal of this kind. Higher-order beliefs are intimately involved in Gricean "nonnatural" meanings. Believing that a speaker intends that we believe that he intends that we believe a particular thing is a fairly complex act of higher-order mental representation. Assigning the correct interpretation to a sentence uttered by a particular person on a particular occasion can be a dauntingly complex enterprise, as we can see from the fact that people don't always understand jokes. But the skills needed for interpretation aren't optional; every member of the community must acquire an adequate version of these skills in order to be considered a competent adult.

If all Lewisian languages have nonnatural meaning, is all nonnatural meaning associated with languages or with conventions more generally? No. Lewis asks us to imagine that someone warns people of the presence of quicksand by making a stick figure out of branches, and then putting it in the quicksand so that it sinks part way down. Obviously, to any passing person, this half-buried stick figure will mean$_{NN}$

"Look out, quicksand!" But it means this not by convention, not because there is some conserved and culturally transmitted precedent that a figure half buried in quicksand means "Quicksand!," not because this is a conventional arbitrary sign. Rather, it means this by sheer force of obviousness, because being shown a mock human slipping down into quicksand makes it pretty obvious that someone is trying to inform us of the danger. This is obvious to you, though, only if you understand that the maker intends you to understand his intent to make it obvious. Otherwise, you'll see only some oddly arranged sticks or, at best, a strangely positioned doll. Then we might say that you didn't *get* it, which, when said of a human adult, is not a compliment.

We might wonder which form of nonnatural meaning—conventional, like the meanings of words, or occasional or "conversational," like the meaning of Grice's insinuating drawing and Lewis's stick figure—came first in the history of the evolution of human communication. The answer isn't as obvious as it might seem. Even a clout on the head or deliberate eye contact can convey a nonnatural meaning of the occasional kind. It can convey an intent to let the person we're hitting, or meeting the eyes of, know that we want him to know he'd better stop what he's doing, or had better not stop what he's doing, or something like that.

(Chimpanzees do hit each other, of course, but probably not to convey this sort of complex multilayered message. They may simply be trying to inflict pain or harm or to take revenge. Direct eye contact is much less common than it is among humans, though bonobos do look directly into each other's eyes during sex. This may be an example of the same cognitive machinery or its evolutionary precursors being used for a somewhat more restricted social purpose. Sex is a rather obvious opportunity for shared attention to a common project, for attending to the fact that the other party involved is attending to what it is we're attending to.)

Consequently, the question is whether we humans started meaning$_{NN}$ things by our words before or after we started meaning$_{NN}$ things by whacking each other on the head or meeting each other's eyes or helpfully pointing at things. We might try to answer this question by looking at what chimpanzees do and seeing if we can find any precursors to either thing. People like Daniel Sperber (2000) and

Tomasello (2008) have made arguments, on the basis of what we now know about the differences between humans and chimpanzees, that seem to imply that some comprehension of "occasional" Gricean non-natural meanings probably predates, and was necessary for, the evolution of our modern human kind of linguistic convention, that something like indicative pointing probably came first. They seem to be saying that it probably arose first as a type of occasional meaning, perhaps in the context of collaborative tool use or the performance of another shared food-gathering or food-processing task, a scenario that would be impossible if the two things were identical.

You can understand indicative pointing only by understanding that the pointer intends that you understand that he wants you to look at the clown or give him the hammer or look under the bucket. When chimpanzees are around humans a lot, they learn to point at things that they want because the gesture may have the magical effect of somehow inducing the humans to give the thing to them. But even when they've learned how to produce it, they are much less likely to be able to *understand* the gesture when it's made by others. They sometimes may be able to understand a human pointing at or reaching for something the human wants, but a disinterested human helpfully pointing out the location of something that the chimpanzee might want appears to be beyond their comprehension.

Chimpanzees can follow a human's gaze, but the human motive of being helpful, of unselfishly managing the attention of others, is apparently so foreign to them that they don't make what seems to us to be the obvious inference that there might be something good in the place that the person is pointing out to them (Tomasello 2008:38–41). Because helpfully pointing out something like that isn't something *they* would do, it seems difficult for them to grasp that we're doing that. They certainly don't seem to have any sense that they're under any *obligation* to figure out what we're pointing at. The whole idea—which seems to play such an important role in allowing us to refer to particular things in the world around us, confident that others will pay attention, and make the effort required to get the reference—is apparently inconceivable. Chimpanzees really do appear to live in the kind of every-person-for-himself world described by the game theorists. And as we'll see, their nonbinding communicative "conventions," though they do seem to have some, are just as sparsely observed as Young's convention of going to lunch at noon.

Unlike the sclera of a chimpanzee's eye, the sclera of a human eye is white, probably because this makes it much easier to follow our gaze and find the object we're pointing to or attending to. Since this is a modification of the person attending, not the person trying to figure out what she's attending to, it suggests that it's been in our reproductive interests for a rather long time that others know what we're attending to, which seems to make sense only if there have been lots of situations in which we were better off because the very fact that we were attending to something could convey to others the suggestion that they, too, should attend to it. Apparently, being helpful in this way, helping others attend to the right things, intervening in the management of attention by others, has been quite helpful to us. This Gricean kind of attention management— letting others see that we'd like to help them see something—is such an important part of human nature that it's actually reflected in our physical appearance.

Dolphins, unlike chimpanzees, seem to understand human indicative pointing (Herman et al. 2000), possibly because they employ the tightly focused beam of sound they use for sonar to illuminate objects for other dolphins. This strikes me as even better than having white sclera. Dolphins also appear to use personal names for one another (King and Janik 2013), another very Gricean behavior, since it requires the user to intend that the party so addressed understand that the user intends that he should attend to her. After many thousands of years of domestication by humans, dogs also seem to be able to obtain information from indicative pointing in some way (Hare and Tomasello 2005; Tomasello 2008:42–43) and can respond to their own names—but they still can't use names to call others, as dolphins can.

With the mention of evolution, we've gone beyond anything David Lewis himself said about human language. We do, however, now have a theory of the evolution of a rather different, more Youngian, nonbinding kind of "linguistic convention," the one articulated by Brian Skyrms in *Signals*. Since researchers like Michael Tomasello and Daniel Sperber seem to have concluded that the kind of Gricean nonnatural meaning that Lewisian conventions have is quite important, while Skyrmsian signaling conventions involve no such thing, we seem to be moving forward from Lewis's achievements in two rather different directions. To sort out this apparent contradiction, I must first make sure that we all know what Skyrms has actually said, so in the next chapter, I will introduce his story about the evolution of his rather different kind of signaling convention.

3

The Evolution
of Signals

SKYRMSIAN SIGNALING CONVENTIONS

David Lewis ([1969] 2002) seems to have two different stories about how conventions manage to remain stable in a human community over time. Often, he seems to say that higher-order rational expectations—my expectation that you will expect that I will expect that you will expect me to go on behaving in the same way—provide the stability. This is also the way that Thomas Schelling (1966) often speaks, and there is an element of truth in that way of talking.

This couldn't, however, be the whole explanation of the stable maintenance of linguistic conventions over long periods of time, or with the involvement of large numbers of people. Misperception is a powerful disruptive force. The attribution, by fallible humans, of expectations and expectations about expectations can be very noisy. As anyone who played the game of telephone as a child knows, an iterated series of small and insignificant changes can add up to a serious distortion of the final outcome.

At other times, however, Lewis stresses that it would be hard for individuals to benefit by unilaterally deviating from convention, from driving on the left while everyone else was driving on the right. A few foreigners who insisted on driving on the left might not be able to persuade a population of people driving on the right to switch to their way of doing things, because the foreigners would immediately have accidents that would discourage them from persisting in their unconventional behavior. A few speakers of Welsh who insisted on using their own language for everyday transactions in London would find that almost nobody understood what they were saying or cared to join them.

The most obvious merit of this second explanation for the persistence of conventions is that it suggests that Lewis's model, or something like it, could easily be made into an evolutionary one. For conventions in general, H. Peyton Young did just that in "The Evolution of Conventions" (1993), though as I mentioned in chapter 2, his kind of conventions differ from Lewis's kind in a very consequential way. Young's games contain no traffic police sanctioning forbidden forms of experimentation and discouraging careless mistakes. Brian Skyrms (1998, 1999, 2000, 2007, 2009a, 2009b, 2010) took the same general approach to signaling conventions, though the same caveat applies to his models.

Skyrms realized that a version of Lewis's simple model of a signaling system could be formalized and turned into an evolutionary game. By doing so, he showed that a population of senders who had a choice, when observing whether the British were coming by land or by sea, of displaying one or two lanterns, and receivers who had a choice, when observing one or two lanterns, of blocking the land or the sea route, could easily converge on one of the two "signaling system" equilibriums: if the British were coming by sea, the sender would show two lanterns, and if the receiver saw two lanterns, the receiver would block the sea route. But if the British were coming by land, the sender would show one lantern, and the receiver would block the land route. Or if the British were coming by sea, the sender would show one lantern, and the receiver would block the sea route. But if they were coming by land, the sender would show two lanterns, and the receiver would block the land route. Any chance fluctuation in the frequencies of the sender's responses to the arrival of the British, and the receiver's responses to the signals, that moved the population even slightly in either of these two directions would be rapidly amplified by the positive payoff to both parties from the receiver's actions, and one of the two signaling conventions would soon become universally accepted.

Here there's no requirement that players "know" anything or have expectations of any kind (let alone higher-order ones) about the other players' behavior. They can be simple automata that randomly choose strategies and are "paid" when the receivers take appropriate action, by having offspring that will follow the same strategy. If successful coordination leads to faster replication, an arbitrary signaling convention will become established in the population anyway. Or the players may blindly imitate other players that have been successful in coordinating with their own partners, leading to a snowballing of imitations of imitations that will take the whole population to one of the two possible conventions. Indeed, as Skyrms argues, simple signaling systems of this kind are ubiquitous in nature. The way bacteria communicate with other bacteria and the way genes coordinate with other genes are perfect examples.

Clearly, the concept of "convention" at work in these cases is considerably pared down even from Lewis's version, which itself pared down what W. V. Quine ([1936] 2004a) thought was required. Technically, in Lewisian terms, Skyrmsian "conventions," if learned by imitation, are imitations rather than conventions, like the unconscious imitations of mannerisms that Lewis discusses in *Convention* ([1969] 2002:118–21). It isn't clear what Lewis would have called the other versions, but the term "Skyrmsian signaling conventions" seems adequate to distinguish them from the phenomenon that he was talking about.

Skyrmsian signaling conventions still are arbitrary and universal, but they need not be "common knowledge" in Lewis's sense, because they need not be known at all. They don't have to have a basis in mutually concordant expectations because nobody involved even has to have any expectations. The signals carry information, but only in the engineering sense of making certain of the receiver's actions more probable. We should remember Bertrand Russell's ([1921] 2010) behaviorist assertion that a person understands the meaning of a word whenever "(a) suitable circumstances make him use it, (b) the hearing of it causes suitable behavior in him" (143).

Quine's original requirement that conventions be both explicit and deliberately entered into has completely vanished, and Plato's paradox about the origins of language has disappeared with it. (Lewis abandoned the requirement of explicitness, and Skyrms dropped the requirement of deliberateness.) This story contains no Gricean attributions of intentions or deliberate human choices, since none seem needed. Signals are sent in response to specific recurring states of affairs in the world, and as in one of Lewis's verbal signaling languages, they simply activate particular contingency plans in the receivers.

I've already mentioned two models of what this blind process of convergence on a signaling equilibrium could consist of. Pairs of senders and receivers who succeed in coordinating might end up having more offspring like themselves, or individuals in the population might observe and imitate the strategies of successful coordinators. These sound like very similar processes, since in each case successful strategies are copied in a way that leaves them more prevalent in the population. They are quite similar, in some ways (Börgers and Sarin 1997). Skyrms uses a simple formal model of this general sort of process: the "replicator dynamic" invented by Peter Taylor and Leo Jonker (1978).

Skyrms also considers the apparently different possibility of learning through reinforcement, the possibility that individuals may start by trying different strategies at random and then increase the frequency with which they try strategies that have produced positive payoffs in the past. To simulate this sort of behavior, he uses two simple, standard models of learning by reinforcement (Bush and Mosteller 1955; Roth and Erev 1995). Because these two kinds of formal model, the replicator dynamic and the models of reinforcement learning, are actually quite similar, Skyrms's evolutionary story and his story about learning are quite similar to each other as well. In the end, it doesn't appear to matter whether we arrive at Skyrmsian signaling conventions by imitation, by trial and error, or through survival of the fittest.

Later, we'll find reasons to doubt that in the real world, these differences really are completely inconsequential. Evolution creates organisms that are innately capable of doing very sophisticated things, but sometimes the same organism also must *learn* to do certain things. There must be some logic to the way in which the work of optimizing behavior is divided up between evolution and learning. Why do some skills evolve and become innate while others are learned? I will address this question in chapter 4, but for now let's just accept Skyrms's way of approaching the subject, assume that the difference is not very important, and see where that takes us.

The model I find most interesting is Skyrms's (2010:118–35) demonstration of the way learning in a population can lead to the introduction of new signals to an existing system. This innovation adds to Lewis's general kind of story an account of the way particular conventions could have come into existence one by one. Once we have that account, we apparently will have one end point of a process by which we might have acquired a language from scratch: Skyrms's model of the evolution of a basic signaling system, of the kind that some animals and even bacteria have, along with what might be a method for proceeding from that beginning to the more complex languages of modern humans by means of an incremental introduction of new vocabulary over time.

We start with something like the "one if by land, two if by sea" scenario. But what would happen if the British chose a third approach—say, arriving by air, for which we would need a new action, such as raising barrage balloons?

The problem we would face at that point, Skyrms notes, is best understood as an "information bottleneck." Although there are more possible situations, and therefore more desirable actions, the number of signals in our signaling system is the same, so there aren't enough signals for one to stand for each state. What we need is a new signal. Any easily repeatable action by the sender that the receiver would notice is potentially available.

In Skyrms's models, the senders try new signals, occasionally and at random, in response to randomly chosen states of affairs. When they first receive a new signal, the receivers randomly choose one of the available plans of action in response. If the action they take isn't the appropriate one, however, their choice will not be reinforced. Only the lucky few who happen to come up with the right response to the new signal—the plan of action that's desirable in the existing situation—will be rewarded, as will those senders who happened to send it to them.

When the new signal happens to be sent in response to a new situation, the right action happens to be taken in response, and both parties are reinforced, the information bottleneck has begun to disappear.

These correlated choices of a new signal to send in response to the new situation and a particular, appropriate action to take in response to the new signal should begin to become more common, in a self-reinforcing process. As they become more common, the advantage they confer will grow larger.

Skyrms formally models this kind of process using what he calls "Pólya-like urns." In a true Pólya-urn process, balls are removed from a container filled with balls of various colors. Each time a ball of a certain color is removed, it's replaced, and another ball of the same color is put into the urn, making some colors more prevalent over time. Sooner or later, by chance, one of the colors will come to predominate through a process of snowballing, and the more common it becomes, the more common it's likely to become.

The urns Skyrms is interested in are merely "Pólya-like" for two reasons. First, he adds one black ball, a "mutator" ball. When it's drawn, it's returned to the urn with another ball of an entirely new color. He also adds positive reinforcement for certain choices. Balls of some colors are returned to the urn with more than one ball of the same color. The consequence is that over time the urn "learns" to produce balls of that color.

Collections of urns that invent new signals, or new responses to signals, and can learn to send them under appropriate circumstances and respond to them in appropriate ways give us a Skyrmsian signaling language that's capable of indefinite expansion as new problems arise in the environment.

We must imagine that the senders have an urn for each situation they encounter, which tells them what signal to send when they encounter it. Every urn contains a black "mutator" ball. The receivers have an urn for each signal they might receive, which tells them what action to take in response to that signal. Each of these urns also contains a mutator ball. Both parties are reinforced whenever the receiver makes an appropriate response to a situation that only the sender can observe. Over time, the senders will learn to send particular, unique signals in response to particular kinds of situations, and the receivers will learn to respond correctly to each signal, because those choices will be reinforced and others will not.

Occasionally a novel signal will be sent in response to a new situation, and in a few of those cases, an appropriate new action will be taken in response. Over time, as those choices by the senders and the receivers are reinforced and other choices are not, the population will learn a new signal, one that reliably elicits an appropriate new response to a new situation.

This is a beautiful model: simple, transparent, precise, revealing, general. Suddenly we understand more about why this kind of process is so common in nature. Surely Skyrms is modeling something real; surely some analogue of this process has led to the many vast populations of signals and signalers we see around us now. A model of such generality and power can explain a great variety of things.

But the model really does seem very general. It apparently applies to bacteria and various sorts of animals and flowering plants and fungi and anything else that's evolving or learning just as well as it does to humans. In fact, all of Skyrms's own examples in *Signals* seem to involve these kinds of nonhuman creatures. We might still legitimately wonder whether there's anything unusual or distinctive about specifically *human* forms of communication. Might there still be a role, in human language in particular, for clauses 4 and 5 of Lewis's definition of a convention? Could that role somehow be connected to the recursive replication of expectations, nonnatural meaning, and all the rest of the distinctive features of human linguistic communication that Lewis, and other philosophers like H. P. Grice, have observed and described?

SOME PUZZLES ABOUT
THE EVOLUTION OF SIGNALS

Now we have our first simple story about the origins of human language. We started out being able to warn of invasion only by land or by sea. But then, faced with the new evolutionary pressure of invasion from the sky, we—without intending to or even being aware we were doing it—blundered our way into a richer, more complex set of signals and responses, merely by signaling and responding at random, in response to the new situation. Then, presumably, the process was repeated again and again until we have the sort of complicated lexicon found in human languages today. This is Democritus's story about the origins of human language, with the hand waving about how we got from the beginning of the process to the present replaced by a rigorous formal model. Of course, Skyrms (2010) doesn't intend this to be a final picture, but it's a good place to start a search for new territory to explore.

At one end of the process, a simple signaling game like "one if by land, two if by sea" becomes established in a population. We know that this can happen because of the many examples in nonhuman nature—from the alarm cries of vervet monkeys to the signals coordinating the activation of different genes in the same cell—of exactly this sort of signaling system evolving over and over. A self-replicating cell differs from a self-replicating crystal mostly in that the labor of self-replication is divided in the cell among many cooperating molecules, all of which differ from one another, whereas in a growing crystal every small group of atoms is nearly the same, so not much coordination is needed or possible. Given the centrality of a division of labor to any form of life, it's hard to see how anyone acquainted with the science of biology as it currently exists could doubt that the evolutionary emergence of this sort of signaling system is both possible and common.

Once the basic system is established, new situations are sometimes encountered to which new responses by the receiver would be appropriate. Occasionally and at random, no matter what the situation is, the senders send new signals they haven't sent before. Sometimes, purely by chance, a new signal is sent for a new situation. Upon receiving the new signal, the receivers engage in random acts. Sometimes, purely by chance, they respond to the new signal with the new act appropriate to the new situation, and both the sender and the receiver are reinforced. The sender's propensity to send that signal in response to that situation increases, and so does the receiver's propensity to perform that act in response to that signal. On other occasions, when the signal is sent in an inappropriate situation or the act taken wasn't the right one, there is no reinforcement. Inevitably, over time the signals become more and more likely to be sent under the right circumstances and responded to correctly. The same general kind of process, as Darwin pointed out in *The Origin of Species* ([1859] 2009), gave us flowers, which survive as signals to pollinators of the availability of nectar.

None of this requires any intelligence or foresight on the part of anyone. Nobody has to have a theory of mind or anything like that. The players can be automata much simpler than any known form of life. To proceed from the vervet monkey's four different alarm calls to the sixty thousand or so words that the average human college graduate might know, apparently all that is needed is this same scenario playing out over and over.

It doesn't seem to matter much how we envision the learning process. It might happen by means of a Pavlovian reinforcement of successful acts of coordination, by adding more balls of the same colors to the sender's and receiver's appropriate urns. It might happen as a result of players imitating anyone who got good results. Or it might happen if offspring inherited from their parents the propensities to use various signals in particular situations and respond to particular signals in particular ways, and successful coordinators had more surviving children. Because of the formal similarity between the models of reinforcement learning that Skyrms uses and his drastically simplified model of evolution (because, according to Börgers and Sarin [1997] and Beggs [2005], the mean-field dynamics of his two learning models end up simply being versions of the replicator dynamic), these three seemingly different mechanisms will produce very similar results.

In fact, this all seems so easy and so independent of any assumptions about the character of the agents involved that it's a little unclear why it doesn't happen *all* the time. What has kept the monkeys using only four signals, or whatever small number they actually use, year after year for millions of years? Why, since they and their similar ancestors have been around for a very long time, aren't they already using sixty thousand words, as human college graduates do? Haven't they ever found themselves in an information bottleneck? Are there really only four different recurring states of affairs in their world? If not, what's the obstacle to coming up with signals for the others, if the monkeys were able to use Skyrms's kind of process, at some point back in the late Eocene, to obtain the first four? How can they *not* have invented any more new signals for tens of millions of years?

If we didn't know any biology, particularly any neurobiology, we could give the easy Cartesian answer to these questions: Well, monkeys are rather simple automata, and no monkey could possibly remember sixty thousand different calls. But we have no such excuse, because we do know that a single cell in the monkey's brain, which must be much simpler than the whole monkey, coordinates its internal activities using tens or hundreds of thousands of molecular signals; and we do know that the monkey's brain contains billions of such neurons, each of which is synapsed with large numbers of other neurons, creating plenty of capacity for their connection weights to store more than four signals. We even know that some monkeys can understand the very different alarm cries of other species of animals living in their environment—for example, the calls of starlings (Cheney and Seyfarth 1990). If monkey cells use tens or hundreds of thousands of signals, and monkey brains contain billions of neurons, what limits monkey societies to a paltry four alarm cries, given that the monkeys demonstrably can learn to understand more and seemingly should be able to do so in an open-ended way by means of some mechanism like the one Skyrms described? Why, in general, are animal communication systems so limited?

And why do humans have such large brains? If no set amount of cognitive complexity is needed to evolve a language and endlessly add words to it, why do the evolution of a large brain and the evolution of a complex language seem to coincide so neatly in human evolutionary history?

Either the two things happened concurrently, or else the large brain came first. Nobody, to my knowledge, has ever advanced a theory of human evolution in which complex languages with tens of thousands of words appeared first, and the expansion of brain size happened later, though now it's beginning to seem to me as if someone should have. Perhaps the australopithecines spoke a very complex language but used it to say only extremely banal things: "Gosh, it's hot today on the savanna!" or "Like termites, do you?" Well, OK, maybe not, but . . . it appears that we need to think about *why* not.

Another puzzling thing about this story is why, among clever creatures like humans, the whole process had to take hundreds of thousands or millions of years to evolve. If blind imitation or Pavlovian reinforcement works just as well as Darwinian evolution as a method of moving from the four-word language to the sixty-thousand-word language, couldn't the whole transition have been completed in a few generations or, at worst, in a few hundred generations? Why did the process need to take hundreds of thousands of human generations?

If material culture is any guide, there seem to have been some extremely long periods of stasis in the recent evolutionary history of humans. For more than a million years, our technology revolved around the Acheulean hand ax. There was an episode of fairly rapid change right before that long period of stability, and another one after it. Why wasn't there always selection for the endless addition of more and more words to our languages, and why wouldn't our constantly more and more complex languages have produced more and more complex cultures? Why did that process ever slow down or cease, and once it did, what could have started it again?

It would be surprising if my extremely simplistic application of Skyrms's very beautiful and general story contained everything we need to know to understand the evolution of actual human languages, in particular. The questions just asked give us something to think about. I'd like to ask one more very general question, and then I need to begin looking at some facts, to see if I can discover anything about the answers to all these questions.

The model I've described appears to have little to tell us directly about the evolution of syntax. A story about adding new words, or even compound signals consisting of several elements, to a lexicon doesn't tell us where the complex internal structure of our sentences might have come from. But if we want to explain where modern human languages come from, we should try to explain that, because it's a prominent feature.

There's another puzzle here. Skyrms's model makes it seem that semantic evolution, the evolution of new words with new meanings, should be the easy problem, while syntax, stringing these words together into more complex combinations, should be the hard thing to master and should have evolved more slowly or later. If that's the right way of looking at things, then what we should expect to see in nature is many signaling systems with a large number of distinct and meaningful signals, but limited syntactic complexity. But if we look at the actual signaling systems of animals like baleen whales and gibbons and nightingales, it's immediately clear that even though their syntax isn't as complicated as that of a human language, it's still much closer to what we can do than the semantics are. As far as we can tell, a nightingale's ability to produce distinct songs greatly exceeds his ability to attach distinct meanings to them, and the same thing seems to be true of humpback whales, all of whom sing the same very complex and slowly changing songs over and over, each year, no matter what's going on around them, and of gibbons, who seem to attach just a few different meanings to their complex, varied calls.

The actual pattern we find in nonhuman nature at the level of complex multicellular organisms seems to be exactly the opposite of the one that Skyrms's very intuitive models suggest that it should be. But why should it be easier to evolve a set of complex rules for musical composition than it is to use the resulting variety to signal various things, which in Skyrms's models seems so easy?

These puzzles should make it obvious that human language can't be understood in isolation. It's one member of a whole natural universe of evolved information-processing systems, but it's an extremely anomalous member of that universe. Brian Skyrms's models seem to have a lot to tell us about what the rest of that universe should be like, but as it turns out, it isn't like that at all. We won't understand why human language deviates so far from what's normal in nature until we understand why the more normal things in its universe are the strange way they apparently are. Consequently, in the next chapter, after considering biological information in general, in the nonhuman part of nature, I'll take a look at birdsong and see what the birds are actually doing with all that semantically empty syntax.

4

Varieties of Biological Information

BIOLOGICAL INFORMATION IN GENERAL

Should we be surprised that there is "information" and "information processing" in biological systems?

I used to think that the spontaneous appearance of something like a computer by chance, simply as a result of atoms and molecules moving around in accordance with the laws of physics, was fantastically unlikely and difficult to explain. But I don't think that's actually true.

Stephen Wolfram (2002) found a universal Turing machine, a computer that can perform, computation that any other computer can perform, that uses only two symbols ("colors") and has only three internal states. This very simple system, with its very few degrees of freedom, can apparently be realized physically in a large variety of ways. Hao Wang (1961, 1965) demonstrated several decades earlier that small sets of square, colored dominoes or "tiles," simple physical objects, can be computationally general in the same way as a Turing machine is.

RNA is already well beyond the complexity threshold needed for computational generality. If you want to make a set of Wang tiles in the form of small RNA molecules, enough distinct sequences with propensities to fold up in particular ways and stick to other specific folded-up sequences can be generated fairly easily. A lab at Princeton University has used such RNA-based Wang tiles to solve chess problems (Faulhammer et al. 2000). As John Conway's Game of Life shows (Gardner 1970), given the possibility of simple information processing, it's very easy for self-replicating arrangements to appear spontaneously. Consequently, one implication of stories like Wang's would, I suppose, be that life must be very common in the universe.

When we say that a set of tiles is "computationally general," one of the things we mean is that it can be set up to self-assemble into any shape at all. On a molecular level, "any shape at all" means almost the same thing as "catalyzing any reaction at all," because many catalysts work by holding together the molecules whose reaction they catalyze in some specific orientation that makes the reaction take place more easily. An infinite variety of self-assembling machines can create an infinite variety of molecular shapes to catalyze an infinite variety of reactions, dividing up the labor of self-replication more finely and carrying it out more efficiently, using more and more specialized tools for each job. So the possibility of evolvable life in our universe is at least partly a consequence of the fact that the property we call "computational generality" is very easy to realize here.

Information processing of fairly complex kinds is fundamental to life. Cells seem to have recognizable memory molecules for storing information, as well as molecules that are used for signaling. There are switches and addresses. There is often machinery for recombining, permuting, or otherwise manipulating stored information in organized and elaborate ways. There is always some machinery or some other provision for filtering out noise. All these are components that any physically realized information-processing device would need.

How should we think about this sort of biological information? Is it a real part of nature, or just something we humans anthropomorphically project onto the brute physical complexity of the real world?

We can think of the sequence space associated with a biological macromolecule like DNA or RNA as a partition of the space of its exact physical states, a division of that full set of states into coarser categories. Each sequence is associated with a very large number of slightly different exact physical states, all of which the cell's machinery would interpret as representing the same sequence. It's reasonable and useful to speak of this partition as genuinely existing in nature because natural selection has optimized cells—which are real, evolved, self-constructing, self-maintaining, and self-replicating natural units—to act as if it does.

Within certain limits, DNA molecules having the same sequence are transcribed into RNA molecules having the same sequence, regardless of the exact physical state of the parent DNA molecule. Everything else about the way the cell is arranged depends on that fact. Of course, if it is folded in the wrong way or otherwise distorted or modified beyond those limits, a molecule may evoke a different response, but as long as it's within these limits, the same sequence will be transcribed in the same way. It is the sequence of the parent molecule, and only the sequence of the parent molecule (not its other physical features), that constrains the gross physical state of the daughter molecule. There is nothing like that going on in a hurricane, even though both a cell and a hurricane are "complex systems."

This insensitivity to the exact physical state of partner molecules creates a kind of simplicity in normal interactions between biomolecules, making their outcome highly regular, even when the molecules themselves are very complex. Biological order, like classical crystalline order, is a simplification and regularization of a complex, irregular molecular chaos, one achieved, in the biological case, at the expense of creating even more disorder elsewhere.

To the extent that a cell's machinery treats two different RNA molecules with the same sequence as if they were identical, it interacts with itself in a way that *simplifies* reality by throwing away some of the available details, responding in almost exactly the same way to many slightly different situations. (As Lewis [(1969) 2002] says we constantly do in using human languages, cells rely on analogies between them.) This is a necessary part of the process by which life continually reimposes its own relative simplicity on the complex world around it. Thermodynamically, the process is a bit like refrigeration, which also expends energy and prevents, or at least delays, the decay of some information, but the simplicity that this process imposes is a much more complicated kind of simplicity. It's an energetically expensive imposition of sameness, a costly filtering out of perturbations and suppression of irregularities. In a cell, "information" isn't something that is *added* to the bare physical particulars of its constituent molecules. Rather, information is those bare particulars themselves, with certain things expensively and repeatedly subtracted from them, a sort of topological invariant that can be preserved through all distortions and translations precisely because it isn't very dependent on exact details. Biological information is whatever makes it through all the various noise filters, including the outermost, very different, filter of the creature's environment, the filter we call "natural selection" (Cloud 2011).

Even genomic sequences must be constantly filtered and repaired by the cells that depend on them. Every free-living organism has some way of repairing damage to its genome, by means of sexual recombination if in no other way. Without this sort of filtration, too much noise would creep in and overwhelm natural selection. Genome repair keeps the mutation rate down to a manageable level. That allows genomes to have more base-pairs, with the same per-genome per-generation mutation rate. This means that organisms can have longer genomes and can pass more information from generation to generation, which means they can become more complicated. One very important constraint on the complexity of any form of life is the sophistication of the noise-filtration systems that it has evolved.

Manfred Eigen (1971, 1992:20; Eigen, McCaskill, and Schuster 1988) showed that an "error catastrophe" will inevitably ensue if the rate of mutation in the transmission of particular traits systematically exceeds the selective disadvantage of the mutated versions. If any one error in transmission is more likely to occur than it is to prevent future transmissions, any finite population will eventually be taken over by the erroneous version; and if any second error in the transmission of that degraded version occurs more frequently than it prevents the transmission of the doubly mutated version, relative to the once-mutated one, it will eventually take over the population. If any third error in the transmission of that even more degraded version occurs more frequently than it prevents the transmission of that version, it, too, will eventually take over the population, resulting in a steady, irreversible deterioration of the original signal into random noise.

I must explain some of the most basic features of Eigen's idea here, if I can do it without getting bogged down in mathematical details that aren't immediately relevant to the subject of human language. The idea of error catastrophe will matter in chapter 5, when we discuss the kinds of "mutations" that can happen to a learned signal, or another item of culture, as it's transmitted between generations, and whether chimpanzees have any adaptations to prevent, control, or repair such mutations. A grossly simplified example should help illustrate the general problem.

Imagine the simplest possible case, a binary genome with only one gene, which mutates to only one mutant version at a rate of 50 percent each generation. If selection against the mutated version doesn't make the fitness of its carriers less than half that of the carriers of the un-mutated version—say they're only two-thirds as fit—and if there are no back-mutations from the degraded version to the original version—which is likely because there usually are more ways of damaging something than there are of fixing it—the population will eventually come to consist only of carriers of the mutated version, because half the un-mutated population will become mutants in every generation, and the countervailing selective process gives the un-mutated remnant only a three-to-two advantage in contributing offspring to the next generation.

Let's start with a hundred individuals. In the first generation, half the offspring will be mutants, so the second generation will consist of fifty mutants and fifty non-mutants. Only two-thirds of the mutants will contribute offspring to the next generation, but half the offspring contributed by the non-mutants also will be mutants, so the next generation will contain 33 + 25 = 58 mutants and 25 non-mutants. Only two-thirds of the mutants in that generation will contribute offspring to the next generation, but half the offspring contributed by non-mutants also will be mutants, so it will consist of 38.66 + 12.5 = 51.16 mutants and 12.5 non-mutants. The next generation will contain six non-mutants, and the one after that, three, and the one after that (rounding down), one. After that, all the non-mutants will be gone from the population, which will continue to decline in number until it has completely disappeared. That, however, is just an artifact of the way I've set up the example; the mutated population could easily persist and subsequently suffer even more mutations.

Once the original un-mutated version of the trait or behavior is lost, if the mutation rate remains equally high and selection remains equally weak, the less-fit mutated versions that are the only ones left will continue to deteriorate even further, so an iterative process will destroy the trait or behavior's adaptiveness fairly quickly. After all, all other things being equal, disorder in the universe tends toward a maximum. There are far more ways of being maladaptive than there are ways of being adaptive, and left to its own devices, with nothing to prevent it, the population will inevitably find its way to one of them.

In the real world, each organism has many traits and many genes. Any sequence on the genome can suffer a mutation, and each mutation occurs with a certain probability, though that probability may fluctuate. Each mutant experiences some degree of positive or negative selection, or is adaptively neutral. A normal population of organisms consists of a cloud of slightly different mutants, what Eigen called a "quasispecies," grouped around a locally optimal "master sequence."

If the overall mutation rate is too low or selection is too stringent, the population will collapse onto the master sequence, making further evolution difficult. A cloud of mutants around the master sequence, constantly being added to by mutation and subtracted from by selection, occurs at a somewhat higher per-genome per-generation mutation rate. This situation is the best one for further evolution, because the many variant sequences in the cloud give the population more distinct chances of finding a way to a beneficial alteration, with the same random mutations taking place against a larger number of different genomic backgrounds. (Remember Lewis's [(1969) 2002] story about the persistence of various slightly different idiolects in a population of speakers of a human language.) But if the mutation rate is too high or the genome is too long, the population will disperse onto the downslope of its adaptive landscape as a result of mutation pressure, moving randomly away from the fittest sequence and eventually dying out in an error catastrophe.

Between too little noise and too much noise is a level of noise that optimizes evolutionary potential, but as the organism and its genome become more and more complex, this level steadily declines. If the genome is a million base-pairs long, a one-in-a-million error rate when copying it is manageable; but if it is a billion base-pairs long, that rate will produce a thousand mutations per genome per generation, which is very different. To sustain the longer genome over evolutionary time, a lower rate may be needed.

Life's solution to this problem has been to repeatedly come up with methods of reducing the rate of mutation and repairing mutated genomes in every generation, with sexual recombination being perhaps the most spectacular. (Although sexual recombination, by itself, doesn't cause the average genome to contain fewer mutations, it does make it more likely that some completely clean copies will be produced in every generation.) As these methods have evolved, genomes have become longer, and organisms have become more complex. When we look at chimpanzees in chapter 5, if we can't find anything that looks like a mechanism for error avoidance or error repair in the transmission of cultural practices, then we might be able to explain the lack of further development in chimpanzee culture as resulting from the inevitability of error catastrophe after the invention and general adoption of any very complex item of culture. Or if we can identify such a defense mechanism, its design may impose a natural limit on the amount and kind of culture that a population of chimpanzees can maintain. At this point, I simply want to alert readers that there is such a problem to be solved, that the potential for error catastrophe is a natural barrier to the intergenerational transfer of complex information, which presumably must be overcome every time that adaptation evolves, in the same way that gravity and air resistance must be overcome every time flight evolves. The physics of information processing creates this problem, which would exist for any sort of organism using any sort of code or language anywhere in this universe or almost any other.

The fact that we ourselves are information processors of this general kind can make it difficult for us to appreciate the full power of natural selection. We often think about information and information processing in ways that reflect the fact that we ourselves are living things, confronted with constraints imposed by our own limits. *We* have to break down computational tasks into pieces and deal with one piece at a time. It takes us time to read a book, just as it takes an RNA polymerase time to produce an error-free transcript of a gene. Many small steps are involved. Whenever we want to do anything complicated, we have to do it in small chunks. Messages must be arranged in a sequential way, one simple piece after another, so that that they can be transmitted and decoded a little bit at a time.

Natural selection has the effect of optimizing designs. When we humans think about solving a design problem, we tend to assume that it will take a certain amount of time or space or a certain number of computational steps to solve a problem of a given complexity. We are very interested in classifying optimization problems with respect to this kind of difficulty. But nature, as selector, doesn't have to respect these classifications, because it isn't an information-processing device solving problems of limited complexity sequentially in stages. Nature the selector doesn't simplify before deciding, nor does it deliberate. Natural selection, the "war of Nature," as Darwin called it at first, happens in the real world, so ideas about computational complexity and bandwidth are irrelevant. Natural selection doesn't care that the problem of protein design is as complicated as the traveling-salesman problem (Pierce and Winfree 2002). It solves it anyway, in no time at all, because it isn't a computer, it's a filter, which lets only properly folded proteins slip through.

A predator may kill its prey in an arbitrarily complicated way, but that doesn't make it any harder or any more time-consuming for the prey to die. The prey doesn't have to use any computational resources or do any work to be affected by the predator's arbitrarily complex strategy, and the event doesn't have to take any particular amount of time. Using terms like *select* and *evaluate* conveys a somewhat misleading impression of the way that natural selection works. Natural selection is not a cognitive or computational process. We are tempted to imagine it as having some of the limitations of such a process, but it does not. Evolution has not made cells what they are today by breaking them down into pieces and evaluating or dealing with the pieces separately, one a time. Although the information in cells is processed serially, it first is produced by natural selection as a gestalt, a single, incredibly complex tangled-up whole. It is the whole organism in all its complexity that either reproduces or fails to reproduce.

The primary optimizing mechanism behind the phenomenon of life simply doesn't care how complex a problem or a solution is. It can evaluate incredibly complex solutions to incredibly complicated problems instantly, and the process doesn't slow down as the thing being evaluated becomes more complicated. Evolution in general and human evolution in particular are very hard to understand without a clear sense of the superhuman power of this outermost filter, natural selection, the filter that filters all the other filters. Natural selection is an oracle for solving certain kinds of complex decision problems in no time at all, like a wind tunnel, not a device of limited power for cheaply computing their solutions, like a computational simulation.

If natural selection is so powerful, why do organisms need complex brains at all? Why can't the oracle of selection simply optimize every organism, in the same way as it has optimized aphid-herding ants or the malaria parasite, to behave in the manner most appropriate to its environment, without a lot of cognitive fooling around in between sensors and effectors? Why do we even have to have a brain in the first place?

The answer is difficult for us to see because this is something our cultural tradition has backward. We tend think of the human mind as the most powerful optimizing force in the universe. That's wrong. No present-day human brain could even design a fly, let alone itself. The truth is that cognition is a weak but cheap way of solving simple but unusual problems that come up occasionally or unpredictably or that must be handled somewhat differently each time they come up. In contrast, natural selection is a powerful but expensive and slow mechanism—requiring thousands and thousands of individuals to live and die over an extended period—for solving problems that are too complicated to be solved cognitively and that occur over and over again in effectively the same way. In a sense, we are evolution's solution to the problem of the sui generis, the unique, the un-categorizable complexity of the real world, in which, as David Lewis ([1969] 2002:36–40) pointed out, complicated events don't ever happen in exactly the same way twice. We're life's solution to the problem that some of the analogies between situations that it has to work with aren't very good and often require a lot of creative interpretation.

In human populations, the power of cultural evolution seems to come partly from the fact that it combines these two advantages. Even though it can be cheap and quick, cultural selection still seems to act like natural selection in that very complex new designs can be evaluated in a few simple steps. You don't have to understand hydrodynamics or do a complicated simulation: just put the boat in the water and see if it floats. If yes, colonize Hawaii. Try the new poison and see if it kills. If yes, use it to hunt with. Plant the seeds from the best ears of corn and see if they'll produce better maize plants. To access this power, living things had to stumble onto a way to make the results of cognition heritable, which took a long time because of the great obstacles to establishing a high-bandwidth, high-fidelity, secure connection between brains. But once we finally managed to find a safe and practical way to do this, we could try out many more different behaviors much more cheaply, and discard or refine ineffective ones without always having to die to do so.

We harness the same powerful force when we domesticate other living things. It allows us to accomplish things we could never begin to understand, as the domesticators of maize, dogs, and cheese-making molds have done. We, in all our unknowable complexity, become the incredibly complex organism's incredibly complex environment, a development that gives us an almost superhuman but rather blind and reckless kind of power. The ancient Egyptians didn't understand the greyhound's genome, or even that it had one. They just kept and bred the dogs they liked, for their own private purposes. Nonetheless, they achieved an alteration in the sequence of that genome that completely reshaped the organism it specified, in ways perfectly suited to their peculiar needs.

BEEHIVES AND BIOLOGICAL INDIVIDUALS

The systems of signals found within particular organisms, in single cells, or going between cells in a multicellular organism tend to be quite complicated. Signals between different organisms, however, tend to be much simpler, with a few very significant exceptions. Bacteria may exchange a few signals with one another, but internally they store and transmit much larger amounts of information. This is also true of wolves and apes. The great apes—gorillas, chimpanzees, and orangutans—strike me as about the worst communicators for their level of complexity that I could imagine. They're apparently the least able to communicate the most complex thoughts. It's interesting, and essential to understanding the glaring exception that we humans represent, to think about why this might be. It can't be that they are simply unable to transmit or respond to very many signals, because they're already doing that internally.

To explain this strange situation, we have to go to the root and ask ourselves what a single biological individual actually is, what counts as a biological individual (Buss 1987). The standard story about this dates back more than a century to August Weissman (1893). The parts of a cell, and the organism of which the cell is a part, make up a single individual organism because they all have to be reproduced or else none of them can be reproduced. None of them can split off on its own and start its own family. A cancer is a dead end, not a distinct organism, because it has nowhere to go, unless it's caused by a carcinogenic virus like the Epstein-Barr virus. The virus can escape and move from person to person, so despite its periodic integration into our genomes, it still behaves like a separate, selfish organism.

Within a single organism, the issue of who benefits never comes up. Either every part benefits equally, in Darwinian terms, or else none does. Consequently, the problem of noncooperative behavior doesn't come up in the same way, either. Signals can generally be trusted, and imperatives should always be obeyed. Parts of the organism that fail to cooperate are destroyed, just as most damaged cells that might become cancers are destroyed before that can happen. Within a single organism, there are equivalents of the traffic police I mentioned in discussing Lewis's ([1969] 2002) example of driving on the right. Attaching the right meanings to signals is in some sense "obligatory," as Lewis and Hilary Putnam (1975b) say it is in a human language. Consequently, evolutionary games of "stag hunt" between different parts of an individual organism don't have to end up at the risk-dominant equilibrium.

This is a very favorable environment for the evolution of Skyrmsian signals. It's a series of pure coordination games in which the same situations and signals occur again and again and must be responded to in exactly the same way by "players" with identical interests. Remember that most of Brian Skyrms's (1998, 1999, 2000, 2007, 2009a, 2009b, 2010) own models of signaling behavior involve pure coordination games, not games whose participants have mixed motives. (The exceptions are the games he uses to show that lying is possible.) Although his models may have quite a lot to tell us about signaling within organisms, it isn't entirely clear that all the results would carry over, unproblematically, to signaling *between* organisms.

When interests differ, when motives may be mixed, it's often in the interest of particular individuals to misrepresent (Dawkins and Krebs 1978). This might not seem like such a serious problem to a human being. We lie all the time, but that doesn't stop us from communicating with one another. But humans are very clever and very good at spotting lies. We make complex recursive representations of what others think, what they must think we think, what they must think we think they think, and so on. All this is very useful for lie detection.

Most animals lack any such sophisticated, generalized ability to discern deception, or any cognitive apparatus they could easily adapt for that purpose. Their only two choices are what we would call extreme gullibility or else an unwillingness to listen at all. A Skyrmsian signaler, which is what all nonhuman, non-cetacean animals seem to be (though I'm not sure about elephants) is extremely, seemingly irremediably, gullible, because it just responds mechanically to a standard signal in a standard way. Effectively detecting lies requires the detector to imagine that perhaps the intention that the liar is representing as being behind his signal is not, in fact, his intention at all. But the Skyrmsian agent never attributed any intentions to anyone in the first place, she merely acquired a conditioned response or inherited a gene. This leaves her with very few resources for spotting liars.

Nature is bad at creating lie detectors, but it is very good at producing sophisticated deceptions, at producing cuttlefish and chameleons and cuckoos (Dawkins and Krebs 1978). Lying is easy—it can even be accomplished through carelessness, by carelessly sending the wrong signal. Spotting one lie in a pile of truths, however, is often quite difficult. You have to find a tiny flaw in the sender's representation, a small feature that isn't exactly right, and then, if the situation is a novel one, figure out why a self-interested lie would have that exact flaw and the truth would not.

The odds in the game of Skyrmsian communication are heavily stacked against the receiver. Anywhere but in a game of perfect coordination, in which deception isn't an issue, the receiver should ignore most signals if she wants to stay healthy. Deafness or complete indifference, all other things being equal, are survival traits. She's too gullible to make it safe for her to trust the other denizens of the bad neighborhood she lives in.

Games of perfect coordination are most likely to appear inside particular organisms, precisely because an organism is a set of parts with identical reproductive interests. Outside these contractual arrangements, only occasionally—when it accidentally happens that the incentives to deceive aren't significant enough to attract enough liars to destroy the game, when lying is too expensive or somehow impossible or the costs of being lied to are minimal, or, for some other exogenous reason, the liars happen to be few enough—will signals be attended to, so usually only in those cases will they be sent. It's inside games of perfect coordination—single organisms—that we find many of the most elaborate signaling systems in nature. But when dealing with other organisms, the problem of deception often makes attending to signals too hazardous.

Some signals—a peacock's colorful, oversized tail as a signal of good health, and a lion's deep, loud roar as an indicator of its size and ferocity— are hard to fake and are relatively safe for a credulous creature to attend to (Dawkins and Krebs 1978). A signal that will certainly be ignored usually isn't worth sending at all. Consequently, in many cases only signals that can't be faked are ever sent, which evidently is a major obstacle to the evolution of more complex forms of communication. This is why, of the relatively few signals sent between organisms, so many of them are expensive and showy displays of some kind. The peacock's tail isn't unusual.

What about warning calls for predators, of the kind that vervet monkeys give? Perhaps responding to a false warning about a nonexistent predator usually isn't very costly in any systematic, persistent way, compared with ignoring a warning call when one actually is merited. With that sort of signal, if individuals occasionally tell the truth, a lot of lying can be tolerated. Apparently, the occasions when sending a false warning is advantageous are often too few to overwhelm the sender's and the receiver's mutual interest, whatever it is, in sending and responding to genuine warnings. But if the warning required a more consequential response, which in the absence of a predator would be harmful or dangerous, and the routinely recurring incentives for lying were substantial, it would be more difficult for a signaling system to remain in existence in the face of ongoing attempts at exploitation by liars. Vervet monkeys seem to have only four things to say, despite being such complex animals, because there are only four regularly recurring situations in which they can actually be trusted (most of the time).

In *The Major Transitions in Evolution* (1998:1–3, 6–10), John Maynard Smith and Eörs Szathmáry argued that in the course of their evolutionary history, living things have had to solve many problems of this same general kind, problems about whom to trust and how to become trustworthy. New methods of transmitting more complex information between generations have repeatedly evolved, making more complicated organisms possible. This greater complexity tends to be achieved fairly suddenly, through the fusion of symbiotic partners, self-duplication and fusion, or some other form of agglomeration:

> One feature is common to many of the transitions: entities that were capable of independent replication before the transition can replicate only as part of a larger whole after it. . . . Given this common feature of the major transitions, there is a common question we can ask of them. Why did not natural selection, acting on entities at the lower level . . . disrupt integration at the higher level (chromosomes, eukaryotic cells, sexual species, multicellular organisms, societies)?   (6, 8)

In many of the cases they discuss, the discovery of a new way of transmitting information has resulted in the symbiotic fusion of formerly competing units, so that units that could reproduce on their own before now can reproduce only as a more complex collective of some kind. Presumably in these cases, we can think of this realignment of reproductive interests as necessary partly to permit the signaling needed to coordinate the new division of labor (which Maynard Smith and Szathmáry point out is also a feature always found in these transitions) in the new, more complex conglomerates. New barriers to entry are often imposed on the new collective to allow its subsumed parts to cooperate and communicate, safe from any intrusion by liars and parasites from the outside world. This is a new, larger game of pure coordination, and around it is a new wall.

Symbiogenesis, the fusion of formerly independent organisms into a single individual (Kozo-Polyanski [1924] 2010; Margulis 1970; Sagan 1967), is one form that this sort of transition can take, but new adaptations like obligatory sexual reproduction may also evolve to allow more intense collaboration between still-distinct organisms. What always seems to be essential is some new system for what Maynard Smith and Szathmáry call "central control," some form of policing, whether it's internal to one new agglomerated organism or (like sexual selection) occurs at the boundary between collaborating organisms. The requirement for "traffic police" that coordination creates is more than just a human idiosyncrasy.

The origin of chromosomes, the origin of eukaryotes, the origin of obligatory sexual reproduction, the origin of multicellular organisms, the origin of animal sociality, and the origin of human language are offered by Maynard Smith and Szathmáry as examples of this process. In each case, the problem of protecting the new conglomeration or association against exploitation by some of its constituent units or parts has appeared and has been, to varying degrees and in various ways, solved. New methods of transmitting information may also require new methods of error correction if longer and more complex messages must be faithfully transmitted from generation to generation.

If the receivers' gullibility is the problem blocking the evolution of more Skyrmsian signaling in nature, then we probably should expect to find fairly elaborate systems of Skyrmsian signals among social insects, whose Darwinian interests are, because of the close kinship of the members of a single colony, nearly identical, despite the relative simplicity of the individual insect compared with an individual ape. And we do. Ants have quite elaborate systems of pheromonal signals, and honeybees also signal to one another in ways that are considerably more complex than those of most organisms (Frisch 1967).

During the "waggle dance"—there is also a "tremble dance" and various other forms of communication—the bees fly in regular patterns whose angle shows the angle between the sun and the direction of the target flower. The number of times they waggle their tails in one part of the dance shows the distance to the target flower. Like words such as *I* and *here*, the dance is "indexical": which patch of flowers a dance picks out depends on when and where it's performed. The angles and distances are always angles and distances relative to "here," wherever that happens to be.

It's been argued that this code had its origins in "practice flights" on the surface of the hive by primitive bees, which were needed to orient the bees relative to the sun before taking off to return to a patch of flowers. Because the practice flights gave some indication of the direction of the flowers, bees were selected to pay attention to them and follow their lead, and the fully developed modern system evolved through a gradual incremental modification of the original behavior. This seems to be the evolution of a simple Skyrmsian convention in a game of almost pure coordination, just like the evolution of signals inside cells.

The existence of this fairly elaborate signaling system in mere insects suggests that the lack of brainpower isn't the obstacle that keeps most other animals from having equally elaborate systems of communication. What a bee can do could also, in principle, be done by a lion, a platypus, a lizard, or a lobster, but it isn't. Instead, what seems to be unusual about honeybees is the extent to which all the interests of the bees in a hive are identical. A beehive isn't quite an organism, but the very close genetic relationship of all the bees in a hive means that the coordination game they're playing is very nearly a pure one. Identity of interests seems to be by far the most favorable state of affairs for these sorts of simple Skyrmsian signals to evolve in, and we see many more elaborate communication systems within organisms, or quasi-organismic kin groups, than systems for communicating between them. When interests aren't almost identical, it seems that communication can easily become confined to signals that are impossible to counterfeit, signals that, for some reason, can't be sent at all unless they're authentic.

BIRDSONG: SYNTAX WITHOUT SEMANTICS

Songbirds seem, at least superficially, to be playing a more complex game. They sing loudly and clearly for everyone, related or not, and the songs they produce, which in many cases are learned, are very complicated.

Since the learning of songs may have evolved independently three distinct times—in songbirds, parrots, and hummingbirds (Catchpole and Slater 2008:77–81)—it seems to have some adaptive value. Birds that learn their songs tend to have local song dialects. Some kinds of birds are mimics (in other words, accomplished liars). Some sing duets with their mates. Indigobirds, which parasitize other birds by laying their eggs in those birds' nests, produce songs that are copies of those of the host species. Individual nightingales can produce hundreds of different songs.

Yet there is still a big gap in complexity between birdsong and human speech. Nowhere in all this variety is there a genuinely recursive, context-sensitive, human-type syntax requiring a computationally general mechanism, the finite equivalent of a Turing machine, a universal computer, to produce it. All the birdsong we know of could be produced by somewhat less powerful pattern-generation machines. It all could be produced by machines with a form of memory that's somewhat simpler than that of a Turing machine, a truly general computer that can do anything any other computer can do.

I pointed out earlier that there are very simple ways of making Turing machines. Remember Dirk Faulhammer and his colleagues' (2000) RNA computer, which was used to solve chess problems. If a small collection of different kinds of molecules can be computationally general, it should be easy for a brain, a much more complicated mechanism, to be computationally general as well. Although linguists like Noam Chomsky ([1957] 2002) have emphasized the unique complexity of human syntax, from a purely mechanical or computational point of view, generating the kinds of strings of sounds that we humans generate just isn't that hard. The right collection of RNA molecules could do the necessary computations.

If generating sequences with our sort of elaborate syntax isn't much more complex than other tasks performed by organisms all the time, then the explanation for the absence of such syntax in the communication systems of nonhuman animals must be functional. Birds must not generate their songs using a computationally general mechanism because they don't *have* to. What birds are actually doing with their songs must be so different from the things humans do with their languages that they simply never need a more complex syntax in the way that we need it.

What is this functional difference? What are the birds doing with their songs that's so different from what we're doing with our languages? What, for example, is the nightingale doing with his hundreds of distinct songs?

What he clearly isn't doing—and this tells us rather a lot—is saying hundreds of different things. Although this may be a slight exaggeration, he's basically saying the same thing in all his songs. It's just more believable, the signal is more credible, the more different ways it's sent.

What is he saying? As the name suggests, the male nightingale sings at night, which is unusual, as most birds sing during the day. Why? Probably because singing at night is extremely hazardous. The nightingale is advertising his exact location to every owl, weasel, and snake in the vicinity, at the very time when he's least likely to be able to see them coming. Singing in this way is a signal of strength, alertness, good hearing, and general fitness that he can't fake, because he has to be able to dodge the predators, night after night, to continue singing for very long.

A male nightingale sings when he has no mate, hoping to attract a female to his territory, so what he's doing is advertising his quality as a spouse. Males accumulate songs over their lifetime, learning new ones all the time, and the variety in a male's repertoire is a signal of how old he is. The older he is, the longer he's already survived this dangerous game. Presumably for this reason, older males are preferred by females, who tend to choose the male with the biggest repertoire of songs. By now, of course—since females with good taste have been having daughters with good taste and sons who sang attractive songs, who attracted females with good taste, who gave them sons who sang attractive songs, for a very, very long time—they probably also like complicated and novel songs, just as humans, by now, like complicated and novel songs, stories, and forms of attire.

Whether or not they do, this case doesn't seem to be an exception to the rule that Skyrmsian signals between organisms must be impossible to fake—in fact, it's a perfect illustration of it. The reason for the complexity of the nightingale's song is basically *cryptographic*; it's to make it hard to copy, which makes it impossible for a young bird that may not survive for very long to sound like a seasoned player. All the syntax isn't there to let the bird say different things—it's there to make the songs complex and difficult to copy.

What about other kinds of birds? They seem to sing for many reasons, but let me give you one more example of the importance of being impossible to fake, one that has special relevance to the evolution of certain aspects of human languages.

Robert Axelrod and William Hamilton (1981) argued that the iterated version of the prisoner's dilemma game could be used to explain some aspects of birds' song learning. The problem in this game is that the players can't trust each other, and so both end up worse off than they would have been if they could. But if the two parties must play the same game against each other over and over again, and if each player can make his own future moves depend on his opponent's past ones in simple ways, they may be able to work out a collaborative arrangement by rewarding cooperation and punishing aggression.

A very important variable in determining how likely this is in any particular case is the length of the expected interaction (Axelrod 1984:126–32). If the two parties will be together indefinitely, it makes sense to try to work out some form of cooperation. But if they're likely to part company soon, the potential for retaliation is limited, and the temptation to be the first to cheat is overwhelming, so they might as well be aggressive.

Many birds learn their songs during a critical period in their first year or so of life. The songs themselves seem to be optimized to be both learnable and as difficult as possible to learn. The repetitiveness, use of pure tones, and other features we perceive as "musical" seem perfect for making it just barely possible to imitate the songs precisely, while their sheer complexity and the virtuosity required to accomplish this make the songs quite difficult to master perfectly. There's some evidence that birds rehearse their songs in their sleep (Dave and Margoliash 2000), so they may have to dream about them in order to get them exactly right.

Learning from adult models during a critical period seems to produce local song dialects, variations in the songs depending on which region the bird is from. Birds also can fly, which means that they can range over wide distances without much impediment. Consequently, they often encounter strangers, birds from far away. A bird with a territory often meets his neighbors with adjacent or nearby territories. He can expect to continue interacting with his neighbors for a long time, so he probably shouldn't be too aggressive toward them. A stranger may be gone tomorrow, though, and has no reason for restraint, so he must be dealt with aggressively at once. (Humans, too, do things to people who don't speak their language, or speak a version of it that shows they're not native speakers, that they would never dream of doing to fellow locals.)

Two birds meeting in the woods are in a sort of prisoner's dilemma of their own, Axelrod and Hamilton (1981) reasoned, since it would be best to be the only one prepared to fight, worse but still acceptable to both to be shy, somewhat bad to have to fight, and terrible to be driven off by a bolder bird. Do you fly toward the other bird or away from him? It would be good for both neighbors if they had some way of telling the difference between birds that grew up locally and birds from outside. Then they could deal with the birds in each category in the appropriate way, acting peacefully toward their neighbors, avoiding repeated costly confrontations, and acting aggressively toward strangers.

Local song dialects, Axelrod and Hamilton conjectured, might be an adaptation to allow them to do exactly that, to tell the difference between locals and strangers so that strangers could be chased away and locals avoided. This, apparently, is an important problem for birds, because song dialects and learned songs are quite common. But this code also can be broken and exploited. Indigobirds, which are nest parasites, may mimic their hosts' songs to lull them into a false sense of security, pretending to be innocent neighbors just passing by.

In both the cases I've described, the complexity of the song's syntax seems to be playing a purely "cryptographic" role, the role of making it just hard enough to learn. The nightingale's song is complicated, so it will take a whole lifetime to acquire a large repertory. Critical periods let young birds learn songs in ways that strangers will come along too late to replicate. What we see in birds, with their relatively small brains, is syntax without semantics, the machinery of permutation without an associated lexicon of interpretations. (I don't mean to claim that parrots, for example, aren't doing something completely different; this is just how I think of the two situations I've described.) Here the function is largely encryption.

Humans, like birds, also seem to have a critical period in childhood. It's harder to learn a new language as an adult than as a child. I believe that this is more of an obstacle to learning new grammar than it is to learning new words. (Nobody would ever make it through medical school otherwise.) Because pidgins are foreign languages learned by unassisted and uncommitted adults, they consist of words strung together without much grammar. Pidgin speakers can learn the words without the modern apparatus of language learning—drills, textbooks, and so on. It's the syntax that's difficult for them to grasp.

If there are languages, like English, without dozens of cases, in which any thought we could express in any human language can be expressed perfectly well, then the dozens of cases in some other human languages would seem to be superfluous. Looking at the complexity of the grammars of human languages, much of which seems to be unnecessary for communication, and considering how hard it is to learn these verbal mannerisms perfectly once the childhood critical period has ended, I can't help thinking that these grammars, too, may exist mostly to make it difficult for adults to learn the language well enough to sound like real natives. At some point in our recent evolutionary history, it might have benefited us, as it does birds, to be able to tell quickly the difference between genuine members of our own tribe and interlopers.

R. M. W. Dixon (1997), who studies Australia's ancient languages, emphasizes the political considerations often behind the development of new languages among modern humans:

> The relationship between two adjoining dialects, or two contiguous languages, is never static—they may be moving closer together in some features and further apart in other ways. Neighboring dialects or languages may gradually converge for a period, and then change direction and begin to diverge. This will be motivated largely by the type of contact between their speakers—friendly or hostile, whether they trade with each other, marry into the other group, take part in sporting or musical carnivals, serve in the army of the other group, and so on.
>
> Once a nation or tribe splits into two, each with its own political organization, the two groups will seize on linguistic features as tokens of self-identification. A handful of lexemes and/or pronouns can be sufficient. The dialects of two new nations or tribes may well be fully intelligible, the important political thing being to take care to use certain words and to avoid others.
>
> Eventually, two dialects may diverge to such an extent that they cease to be intelligible and must be considered distinct languages.   (58)

Because these sorts of differences in language play such a central role in structuring social interactions between distinct human social and cultural groups, Dixon (1997:63–66) himself doubts that human language could have evolved gradually. He thinks that simpler languages would have been too easy to learn and couldn't have played this isolating role successfully. A less radical model would be one in which as our ability to handle syntax slowly evolved, our languages stayed just at the very edge of learnability, like birdsong, an edge that kept moving outward, slowly or rapidly, as our brains and behavior changed.

The strange and apparently superfluous complexity of modern human languages, whose grammars may take hundreds of pages to describe accurately, would then be the result of a long arms race (Dawkins and Krebs 1979), in which some members of a population were continually being selected for their ability to learn, as children, to speak a language that it would be very hard for adults to learn perfectly, while other members of the same interbreeding population were being selected for their ability to learn languages, as adults, that ought to be very difficult for them to master once the critical period was over.

Dixon (1997:22–23) points out that the speakers of a "prestige" language seldom bother to learn less prestigious languages, whereas speakers of less prestigious languages often learn the prestige language. Making this harder presumably makes the club of speakers of the prestige language more exclusive, but being excluded from that club probably hasn't been particularly good for the reproductive fortunes of those who couldn't get over the hurdle.

A similar story could probably be told about pronunciation, in any language—the tones of a tonal language like Thai or Yoruba, or the clicks of a click language like !Kung, or all the various British accents. If you don't learn these things during the critical period, it takes a lot of work to get them exactly right, and people will judge you instantly by how well you've succeeded. So nearly everything I've said about birdsong may apply almost equally well to some aspects of the syntax and pronunciation of human languages, though not to their semantics.

Even if that rather cynical thought is true, however, a lot of other things are going on in a human language as well, things that are not about encryption or membership in an in-group. Lying is quite easy in any human language, so our signals are not impossible to fake, unlike the peacock's tail. The question is how we've managed to overcome the problems with trust that keep the signaling systems of most animals in such a rudimentary state. Perhaps a look at the chimpanzee, our closest living relative, who seems to be right on the other side of this barrier from us, will shed a little light on what might have happened in our particular case.

5

The Strange
Case of the
Chimpanzee

SIGNALS WITHOUT SYNTAX

We have a tendency to look at apes as if they were rudimentary humans. We try to understand their communicative abilities by asking to what extent they're capable of mastering *our* languages. The results of these inquiries are somewhat controversial (Savage-Rumbaugh et al. 1986; Seidenberg and Petitto 1987), but they have shown two things fairly unambiguously. Chimpanzees aren't remotely capable of mastering the full complexity of a natural human language, whether spoken or signed. If properly taught, however, they are capable of understanding and employing a much larger number of signals than wild chimpanzees use.

This means that chimpanzees present us with exactly the same Skyrmsian puzzle as the other organisms I've discussed. If they're cognitively capable of understanding more distinct signals than they actually use in the wild, why don't they invent some new ones? Why have they been just sitting around for millions of years not inventing a more complex language while our ancestors did the opposite?

We might think of this mystery as linked to a similar mystery about chimpanzees' use of tools. Since chimpanzees make and use tools, up to and including pointed sticks or crude spears (which they've been observed using to hunt galagos [Gibbons 2007]), and groups in different places use different tools, indicating some sort of cultural transmission of tool use and/or toolmaking practices, why have they never gone beyond breaking open nuts with hammerstones and stripping protruding branches off a sharp stick? Why have they never learned to make simple stone tools by knocking chips off a cobble? Hitting a nut with a stone and hitting a stone with a stone don't seem that different, so why have chimpanzees been doing the first thing for millions of years, without ever being tempted to try the second?

Not only is that implausible, but it probably isn't true. We have seen captive bonobos and capuchins smashing stones to get sharp flakes and then using the flakes to cut with (Toth et al. 1993; Westergaard and Suomi 1995). That, however, only deepens the mystery. Why, having learned to do that, as they might have many times during the last few million years, would a group of chimpanzees ever forget that knowledge? Why wouldn't they remember the technique and start endlessly accumulating improvements?

With respect to their communicative abilities, the simplest available theory is that they, like other animals, are, by human standards, very gullible but not very trustworthy and so are better off not attending to Skyrmsian signals, unless those signals cannot be faked or are foolproof in some other way. If that's the right explanation, then we should expect them to produce more or less the same kinds of signals as other animals do. These signals should either be difficult to copy (like the songbird's dialect, the nightingale's large repertoire, or the peacock's hard-to-maintain tail) or usually benefit both parties (like the growl that makes a fight unnecessary). Or, like predator warning calls, ignoring them should usually be dangerous when they're true, and attending to them should not be very costly if they're false.

Chimpanzees do have a set of signals that more or less fit that description, screams of fear and cries of excitement. The vocal signals of wild chimpanzees are relatively few and seem to be largely innate and involuntary, like laughter or screaming in humans. Of course, this doesn't mean that they don't contain a lot of information just in the "tone of voice," as a laugh or an inarticulate cry can in humans.

If the fewness of the calls is a result of the fact that there aren't many situations in which a chimpanzee can be trusted, there's little reason to expect that the amount of information actually conveyed on those few occasions will be small. It isn't cognitive complexity that's limiting communication, it's trust, so signals sent when the sender is likely to be trustworthy may be fairly complex and informative. On such occasions, telling others a lot is at least theoretically possible; what should be impossible is ever saying anything new. The premium is on economy of expression rather than on flexibility. A scream of rage or a sob conveys its message as a holographic gestalt, because there's no need for the kind of modularity that would allow infinitely many arbitrarily different messages to be sent over the same channel.

In this sense, these vocal signals are typical Skyrmsian signals; nobody has to intend anything to send them, and no attribution of intent is necessary to interpret them. They indicate the sender's attitude, or emotional or physical state, when there's a recurring and reliable common interest in having that known, like a wolf's snarl. Only a few standard situations can be reported, so they share the decidedly finite character of a Lewisian or Skyrmsian "signaling language," like the language in which two lanterns in the window mean one thing and one lantern means the other.

Chimpanzees don't even seem to have the kind of minimal "referential" vocal signals for indicating the presence of a predator that some monkeys have. In fact, in the context of other complex organisms, chimpanzees and other great apes are not conspicuously good at vocal communication. Gibbons sing, but chimpanzees don't. Elephants may be better communicators. Apes' cleverness may make them especially untrustworthy, or there may be some other explanation, but for whatever reason, chimpanzees seem to combine an unusually sophisticated ability to model one another's minds with a remarkably poor capacity for conveying to one another what's in them. If a Martian had visited Earth 7 million years ago and been asked to guess which of the planet's animals would eventually evolve the most complicated system of communication, he probably would have unhesitatingly picked dolphins over the common ancestor of humans and chimpanzees.

(Perhaps I'm imagining an unsophisticated Martian; a more experienced one might know that it's precisely this sort of bottleneck that is likely to elicit the evolution of a completely new system to overcome it. He might see dolphins as permanently stuck with a pretty good solution and therefore unlikely to evolve the kind of radical innovation needed to start a full-blown process of human-style cultural evolution. We won't know what he should have expected until we have more examples of cumulative cultural evolution to study. We also need to learn more about dolphins.)

What complicates the situation with chimpanzees, however, what does seem to give them some evolutionary potential as communicators, despite their rather rudimentary system of vocalizations, is the existence of a second, separate, learned Skyrmsian signaling system. Chimpanzees also communicate with gestures, gestures like holding up a cupped hand to beg, raising one arm (as if the individual is about to playfully strike a peer) to initiate play, or deliberately ripping a dry leaf to attract attention.

The existence of this second system of learned signals coexisting with the first system of innate signals should interest us because the project of this part of this book is basically an attempt to explain one of John Maynard Smith and Eörs Szathmáry's (1998) major transitions in evolution, which are supposed to always revolve around a new way of transmitting information between generations. Here, right before our eyes, in our closest living relative, we have a new way of transmitting information in what appears to be a rather undeveloped form. The coincidence seems suspicious.

In chapter 6 of *The Origin of Species* ([1859] 2009), Darwin mentions redundancy as a circumstance favorable for the further evolution of an organ or a system, since it means that selection on one of the two redundant versions can be relaxed, allowing it to vary: "Two very distinct organs having performed at the same time the same function, the one having been perfected whilst aided by the other, must often have largely facilitated transitions" (186). "Duplication and divergence" has become one of the central mantras of modern evolutionary theory. In chimpanzees and other great apes, the existence of two systems with very similar functions, two distinct systems of Skyrmsian signals, one innate and the other learned, might be thought of as creating a

Darwinian redundancy that would permit the function of one of the two to diverge in an unexpected direction. The second system of signals seems like exactly the sort of thing that evolution might easily have grabbed hold of, in a descendant of the common ancestor of humans and chimpanzees, and turned into something new, into a more human kind of language. If that's what happened, then understanding the two end points of the process is the best way of understanding the transition, and that makes understanding what chimpanzees do with their gestures rather urgent. So let me tell you a little bit about it.

In writing this chapter, I relied heavily on the work of Josep Call and Michael Tomasello (2007; Tomasello 2008; Tomasello, Call, et al. 1994, 1997; Tomasello, George, et al. 1985), especially their account of chimpanzee gestures in "The Gestural Repertoire of Chimpanzees," though any mistakes in my interpretation are, of course, my own. The idea that apes' gestures are a logical starting point for the evolution of a more complex language is one that Tomasello and Call, and others like David Armstrong (2008), have favored for a long time.

There seem to be about thirty or forty distinct gestures that are sometimes used by some chimpanzees for communication. When the number of gestures that a particular chimpanzee uses reaches its peak, he or she may have a repertoire of around twenty communicative gestures (Call and Tomasello 2007:29–30; Tomasello, Call, et al. 1997). Some of the less common gestures are found in some groups of chimpanzees but appear to be absent in others, making it conceivable that, as with human languages, at least some cultural transmission is necessary for their acquisition (Call and Tomasello 2007:33). Despite these similarities to human language, however, there also are pronounced differences.

There's some difference between groups of chimpanzees in the gestures their members make; but there is also nearly as much difference, within each group, in the signals that each member uses. One study found that 14 percent of the gestures observed were unique to one individual (Tomasello, Call, et al. 1994). At any given time, 40 percent of the gestures used by the individuals in the groups studied were unique to particular individuals in those groups. These communicative conventions are rather sparsely observed, it appears, like the nonbinding convention of going to lunch at noon.

No single chimpanzee uses all the gestures used by chimpanzees in her group; each one uses an idiosyncratic sample of the full "lexicon." Members of the same age cohort are closest to one another in shared vocabulary, while greater differences in age produce greater divergence. (Some signals typically drop out of an individual's vocabulary as he matures, while others are added, but this seems to explain only part of the difference between cohorts.)

These facts seem to reflect the strange way that the gestures are learned. Unlike humans, chimpanzees apparently don't learn to communicate by observing the communicative behavior of the adults in their community and attempting to reproduce it. Instead, they make the gestures spontaneously—for example, during play—and they become "ritualized" through a series of increasingly stereotyped repetitions. Gestures that begin specific sorts of interactions eventually become signals for initiating those interactions, with an exaggerated version of the first step becoming an invitation to respond with the next. Animal behaviorists call this exaggerated initial step, which often is part of systems of innate signals as well, an "intention movement" (Tomasello 2008:22–26). A good example of a learned intention movement is the begging hand, which begins the task of accepting what is hoped will be an offered nut or piece of fruit. Other gestures, like "leaf clipping," loudly ripping a leaf, seem to be attention getters.

That the easiest way for a chimpanzee to attract another chimpanzee's attention to something may be ripping a leaf in its vicinity, rather than pointing at it, tells us a lot about the differences in the kinds of things that humans and chimpanzees habitually attend to. We're obsessed with our unending cooperative project of paying attention to the same thing the people around us are attending to, and are constantly trying to follow their gaze. In interpreting pointing, we presume that the human pointer is trying to be helpful to us in that shared project. Again, this appears to be why the sclera of a human eye is white; it makes following our gaze much easier.

Chimpanzees would seem to be paying more attention to finding out what's going on around them in the forest. They sometimes try to follow the gaze of other chimpanzees, but without getting much help from the individual whose gaze is being followed. Unlike dogs, who may have been modified in this respect as a result of their domestication by humans, they don't find it easy to understand indicative pointing, though they do gather more or less the same kind of information by observing perceived rivals reaching for things, apparently because it's very difficult for them to grasp the idea that the pointer is trying to be helpful (Tomasello 2008:38–41). Their way of conceiving of others' intentions doesn't include helpfulness as a possible motive. They are automatically in competition with anyone who isn't their mother or, if they're a mother, their child. Consequently, when a chimpanzee in the wild wants to direct another chimpanzee's attention to something, his sexually aroused state or involuntary grimace of anger, he must resort to the more manipulative expedient of deliberately making an alarming sound.

While this is different from what humans do, because of the lack of any assumption that there are common interests, Tomasello (2008:22–29) points out that it is an example of one individual deliberately manipulating the attention of another individual. This seems to put it very close to the kind of mutual attention-management typical of humans. Nevertheless, the lack of any perception of common interests would appear to put Lewisian conventions—which depend on everyone in the community having, perceiving themselves as having, and perceiving others as having a common interest in coordinating their behavior around the precedent on which the convention is based—permanently out of their reach. An utter inability to trust others or even to conceive of trusting them, an inability that makes it literally *unimaginable* that they might helpfully point something out, is a bad basis for coordination built around concordant higher-order expectations and the presumption of shared goals. Apparently nothing in human social life is possible in the absence of some mutual expectation of goodwill.

How does this very unfamiliar kind of learning process result in differences in the signal repertoire of different groups, and different age cohorts within groups? Since individuals in the same age cohort tend to play with one another, it would not be surprising if they all encountered repetitions of the same gesture, as the individual who invented it used it on various playmates. The widespread comprehension of the gesture that this repetition might eventually produce would make it easier for other individuals who came up with the same gesture independently to elicit a good response. Some of the cultural transmission involved, if there is any, therefore might be indirect, by way of the audience, rather than from senders imitating other senders. Audiences that already respond to a gesture in a predictable manner should be fertile ground for its reinvention. This would make the audience, and not the signaler, the bearer of chimpanzee signaling culture, such as it is. Each new signaler must reinvent the signal for an audience whose interpretations already may be partly fixed.

Why *aren't* the gestures learned by imitation? Unlike humans, chimpanzees don't engage in role-reversal imitation, or at least they don't do it very often or very willingly. The difference seems to be at least partly dispositional rather than cognitive. They can be *trained* to imitate on command (Custance, Whiten, and Bard 1995; Hayes and Hayes 1952), so they're obviously capable of it, but usually they don't seem to want to. There is some evidence that chimpanzees will imitate an action when there seems to be no possibility of figuring out for themselves how another agent's performance of that action has produced an immediately desirable result (Horner and Whiten 2005). But most of the time, imitation seems to be beneath them. Certainly they would never do it without a motive, without a visible reinforcement like food waiting for them if they could only overcome their reluctance.

There is a lot of evidence that role-reversal imitation isn't how chimpanzees learn communicative gestures, that instead they learn them by means of "ontogenic ritualization," by the gesture occurring spontaneously in play or another interaction and then gradually becoming stereotyped as it's repeated in the same situation (Call and Tomasello 2007:5–6; Tomasello 1996; Tomasello, Call, et al. 1994, 1997; Tomasello, George, et al. 1985). The interesting question is why there is such a difference, why it's optimal, or at least feasible, for chimpanzees to learn to signal in one way but optimal for humans to do it in another. It can't be the absolute difficulty of imitation, because some birds imitate beautifully, despite having much smaller brains. (Do we really think that parrots are smarter than chimpanzees?) It's almost as if there has been selection *against* imitation.

Members of the same age cohort may surreptitiously imitate one another's gestures to some extent, though there's little evidence of that. Primates apparently have neurological adaptations, "mirror neurons," for reproducing hand motions that they see others making, with the chimpanzee homologues of Broca's and Wernicke's areas—which in humans are essential to the use of language—among the areas that include these neurons (Rizzolatti et al. 1996; Taglialatela et al. 2008). Even if they do surreptitiously imitate one another sometimes, what should strike you as unfamiliar about this story is how *little* human-style imitation of the communicative acts of others there is, how poor the transmission of vocabulary between generations is, how unlike a human language this whole system is. In its own way, what the chimpanzees do is very clever, very creative, and very impressive—but it is completely different from what we do.

The question is what's different and what is the same, and why. Some differences are obvious. Even if chimpanzees do imitate sometimes, they obviously do it a lot less than humans. Chimpanzees seem as reluctant to imitate as we are eager. They appear to be better at forgetting items of vocabulary. Probably the most striking differences are the presence of idiosyncratic signals used only by certain individuals, and the fact that not every adult member of the group can use every signal used by every other member. Although individual humans do have idiosyncratic vocabularies, all competent speakers use the core vocabulary of a human language. This simply isn't the case for chimpanzees, even though the number of signals in use is much smaller.

This seems remarkably haphazard. We normally think of Darwinian evolution as producing finely optimized adaptations. If the ability to signal with gestures is adaptive for a chimpanzee, surely it would be better to be able to recognize and use all the signals used by anyone in the group. And why should they have two distinct systems of Skyrmsian signals, one innate and the other learned, in the first place? What is the difference in the two systems' roles in chimpanzee life, and what is each used for?

Chimpanzees aren't unique in this regard. Songbirds that learn their songs, as the nightingale does, also have other cries, such as alarm calls, that aren't learned. Here, the reason for the difference is obviously functional. In normal communicative situations, songbirds can use innate calls because they don't need any sort of encryption to make the signal foolproof, but if mimicry might become a problem, learning is necessary. This, however, can't be the right explanation for the two distinct chimpanzee systems of communication, because there's no reason to believe that gestures play a role in recognizing who is and who isn't a member of the local group, or anything like that. There must be a difference of function, but it's hard to believe that it's the same difference.

THE FREE PLAY OF SELECTIVELY INCONSEQUENTIAL BEHAVIORS

The real difference, as far as one can be identified from the evidence, seems to be that innate calls are used in situations in which it might actually matter, from the point of view of an individual's fitness, whether or not he communicated something to the chimpanzees around him. The distinctive feature of the situations in which learned gestures are employed, or at least the aspects of those situations they're used to influence, is that the ability to use the gesture often doesn't seem to matter very much to an individual's fitness.

In the cases that Tomasello and Call (Tomasello, Call, et al. 1994, 1997) studied, between 47 and 70 percent of the gestures were used during play. The rest were used in various other contexts, including "affiliation, antagonistic, feeding and nursing, sexual, grooming, travel" (Tomasello and Call 2007:30). This heavy bias toward play as a context of use raises the intriguing possibility that the system of learned signals may exist as a mere side effect of the chimpanzee's general intelligence, not as an evolved adaptation, that this behavior isn't something chimpanzees couldn't survive without.

Evolving an elaborate set of signaling behaviors just for play would be odd. Perhaps chimpanzees' general shrewdness and skepticism have the side effect of allowing them to learn and employ Skyrmsian signals, by means of something like the semiaccidental method that Brian Skyrms (2010) modeled, despite the absence of identical interests in many of the situations they encounter. In that case, the perceived common interest that motivates much of human communication would ultimately be descended from a mere willingness to play together, when there was nothing at stake.

If a big, dominant male chimpanzee is in a rage, it may be in his reproductive interest, and in the interest of everyone around him, for that to be known. When two subgroups are near each other in the forest, it may be in the interest of everyone involved for the groups to meet. The innate vocalizations convey these indispensable messages, and lives are saved. But if one young individual wants another young individual to play with him or groom him, he will probably be able to convey that information somehow, sooner or later, even without a Skyrmsian signaling convention to help him. Drawing attention to oneself by ripping a dry leaf may be a way to attract attention in preparation for sex—but what probably determines how much sex a male chimpanzee can have is his position in the dominance hierarchy, or something like that, not how adept he is at drawing attention to his readiness, so the action seems somewhat superfluous. There must be other ways of accomplishing the same thing, since not every male chimpanzee masters this gesture.

The fact that not every chimpanzee knows how to make every gesture, or even every gesture current in his or her own group, suggests that it can't be vital to fitness to know them. Why trust the acquisition of an important behavior to such a haphazard and inefficient system? These are all reasons for considering the possibility that the whole system of gestures may be what Stephen Jay Gould and Richard Lewontein (1979) called a "spandrel," an accidental side effect of selection for other features—in this case, for general cleverness and a basic theory of mind—that may or may not subsequently acquire new adaptive functions of its own. In humans, it eventually did, but in chimpanzees, according to this theory, it has remained a useless but harmless spandrel.

If the gestures are of minor importance to fitness, why make them at all? The most obvious answer comes from Skyrms (2010). If any creature whatsoever that was good at learning by means of reinforcement would probably invent a system of signals in a game of pure coordination, then chimpanzees could easily invent a system of signals covering those of their interactions that are games of pure coordination. Since situations, or details of situations, that don't affect fitness also don't involve conflicts of reproductive interest, the existence of any other sort of reward can easily make them into games of pure coordination. If specifically psychological rewards are available to chimpanzees in selectively inconsequential situations like play, they *should* invent a biologically useless system of signals to help them coordinate in achieving them, because they have a motive and the means, and there's nothing to stop them.

Selectively inconsequential behavior is behavior that has equally good alternatives. It's a requirement of a Lewisian convention that there be some alternative that's almost equally good for everyone involved (Lewis [1969] 2002:68–76). Although we could use two lanterns to indicate invasion by land, we in fact use one. This similarity between the two end points of the evolutionary process that produced human language—with both ends apparently involving behaviors with perfectly good alternatives—is intriguing.

The sexuality of bonobos is another example, in a very closely related animal, of the way in which aspects of a behavior that initially are selectively inconsequential can eventually acquire an adaptive function of their own. In every generation, in a group of chimpanzees, the males and females have to accomplish a certain number of matings, but exactly how those individuals mate, or what other sexual interactions members of the group may or may not have, doesn't matter directly to fitness. Chimpanzee populations might be imagined as running through all the available alternatives in an adaptively neutral way. This random process might have continued until something like bonobo sexuality emerged by accident as a result of this very clever animal's quest for psychological pleasure. But the close ties between females that resulted from the bonobos' adaptively inconsequential sexual interactions apparently turned out to be so useful for keeping things peaceful, for collectively discouraging violence by individual males, and for initiating a process of "self-domestication" (Fruth and Hohmann 2002; Hare, Wobber, and Wrangham 2012; Hohmann and Fruth 2003a, 2003b) that they were preserved by selection in a local population that then became the ancestors of modern bonobos.

(When gangs of female bonobos punish excessively violent males by beating them up when they misbehave, we see a pursuit of a shared goal—peace and quiet—which looks very much like a precursor of the social contract. If Jane Goodall had chosen a population of bonobos to study, instead of the more violent and fractious common chimpanzee, we might not find the ability of our chimpanzee-like ancestors to evolve into a creature capable of sharing goals and helping one another quite so mysterious.)

The fact that two of the three kinds of descendants of the common ancestor of humans and chimpanzees have elaborate adaptations that seem to have emerged out of the free play of selectively inconsequential behaviors (bonobos have bonobo sexuality, and we have human language) suggests that this is simply a potential that exists in our general sort of creature. The playfulness and cleverness of the common ancestor of humans and chimpanzees were available to do whatever job nature could find for them, and in our two divergent lineages, they found two rather different uses.

In this story, the evolved signaling system is the innate one, while the learned signals exist as a sort of side effect of the animal's general intelligence. Chimpanzees are extraordinarily clever animals. They exploit many different scattered and inconstant sources of food—different kinds of trees with different fruits or nuts in different seasons, as well as things like termites and monkey meat. Their cleverness even extends to inventing and employing simple tools, using hammerstones and hunting with sharp sticks. In each generation, in the same way that they invent tools, they might just as easily invent gestures to get what they want from other chimpanzees, but only in behavioral domains in which there's no selection against it, in which no evolved unwillingness to attend to signals exists. We might expect to find such signals used in contexts like sex, grooming, and play, in which individuals are engaged in forms of coordination that offer psychological rewards but whose exact details are relatively inconsequential, provided the activities take place *somehow*. This is, in fact, where they're used, so to that extent, theory and reality match.

But if chimpanzees possess these kinds of learned or invented Skyrmsian signaling conventions, the mystery only intensifies. How can they have just sat there with only twenty or thirty of them, for millions of years? Why haven't the inventions accumulated into the thousands by now? Chimpanzees' behavior—the recurrent process of "ontogenic ritualization"—shows that new Skyrmsian signaling conventions do arise, over and over, in each new generation. The question is why they don't persist and pile up.

One reason is obvious, though it's also fairly puzzling. Despite engaging in various other complex interactions, despite having fairly shrewd ideas about what other chimpanzees want or how they would respond to some prompting, chimpanzees apparently are much less prone than humans are to immediately imitate or echo the communicative acts they witness. Chimpanzees find their role in a ritual in a way that's quite different from the way that humans find their role in a convention, and this difference affects the nature of the whole system.

Chimpanzees' toolmaking has the same weirdly un-imitative character. Observers agree that chimpanzees don't learn to make tools by imitation, or at least are much worse at it than human beings are, and imitate much less often. In fact, it isn't clear that wild chimpanzees learn by imitation at all. Some researchers argue that they don't, and others claim that they do, a little. Certainly we have evidence that they can imitate a little in unnatural, experimental conditions, or when adopted by a human family (Custance, Whiten, and Bard 1995; Hayes and Hayes 1952).

Again, even parrots can imitate some fairly complex behaviors, so it isn't obvious that it's the cognitive complexity of imitation that prevents chimpanzees from doing it more often. The difference seems to be at least partly motivational. Apparently, it just doesn't occur to chimpanzees to do as much imitating as we do. It isn't something they particularly want to do, any more than they naturally want to be helpful, as humans do. If they are immersed in a society of helpful and imitative humans, chimpanzees seem to be able to acquire these dispositions to a limited extent, but the behaviors aren't very natural for them, and they remain quite difficult.

Much of what chimpanzees learn from being around others, they instead learn by what observers call "emulation" (Call and Carpenter 2003; Tomasello 1996; Whiten et al. 2004): "I know he did something with that twig, and I know he got termites, and I like termites, so, hmm, let's see, what could you do with this twig that might get you termites?" A mother often shares the fruits of her tool use with a child, creating a distracting incentive to focus on what she could be doing to get them. She may not resist when her child steals her hammer; she may just go find a new one (Boesch 1990). But the direct imitation of whole complex sequences of actions seems to come less easily to chimpanzees than it does to human children, and the individual being imitated or emulated makes far fewer concessions to the emulator's convenience.

The question is why. The assumption sometimes seems to be that mimicry is too difficult, but parrots' and chimpanzees' ability to imitate when rewarded for doing so by humans seems to prove otherwise. This has traditionally been treated as a psychological question, with the need for the human capacity for imitation assumed and the cognitive or behavioral obstacles then enumerated. The chimpanzee is an unfinished human being; what would perfect him? That can't be the right way of looking at things, however, because no organism is ever evolving toward any distant goal. At any given moment, it is an almost perfect whatever it is—an almost perfect flying squirrel, not a rudimentary bat. Otherwise it would be extinct. What must instead be true is that emulation, not imitation, is the *right* way for chimpanzees to learn to use tools and to signal. But why?

WHAT KIND OF "MAJOR TRANSITION"?

This whole situation seems quite puzzling. If we look at the chimpanzee as being on one side of a "major transition," however, with ourselves on the other, it doesn't seem completely inexplicable. According to Maynard Smith and Szathmáry (1998), the general pattern of major transitions is that as the result of an innovation in the way information is passed between generations, and the possibility for more complex evolved systems that this creates, formerly independent units become irreversibly interdependent for their survival and reproduction, aligning their interests to an unprecedented extent. The danger that the new cooperative arrangements will be exploited means that new barriers to entry, new forms of encryption, or other defensive arrangements are likely to become necessary.

In the origin of sociality in insects, the obstacle was the clashing reproductive interests of the individuals involved. This was overcome by finding ways to make the individuals in each social unit very closely related, so that raising the offspring of a single queen would be in everyone's reproductive interest. Sometimes we seem to assume that the main obstacle to the evolution of human societies must have been the same: once again, the animals involved had to become altruistic, to subordinate their reproductive interests to the good of the group. This, however, is not the only possible hypothesis. Each of Maynard Smith and Szathmáry's major transitions involved rather different obstacles. In the human case, another, slightly different, or more complicated set of qualities might have been required.

Instead of simple altruism, as Robert Boyd and Peter Richerson (1985, 1996, 2005; Richerson and Boyd 2004) have long argued and as Kim Sterelny (2012) has maintained more recently, we might want to look to something that really is unique about humans: our relationship with culture. This suggests a rather different picture of the major transition with which we're concerned. Clashes among biological individuals might not have been the only problem that needed to be resolved. In addition, the new system of inheritance—human cultural learning—must have created its own new risks.

The question we have to ask ourselves, in thinking about the nature of this major transition, is what would happen to a slightly more imitative population of chimpanzees. What if they were just a little more interested in observing and imitating one another's activities, and a little less focused on emulation, but in every other way were exactly as we see them today?

As ways of making and using tools and sending signals continued to be invented at the same rate but started to be copied with more enthusiasm, they presumably would begin to accumulate. However, there's no sign at all in chimpanzees of any adaptation for ensuring that cultural behaviors are passed on faithfully. In fact, the contrary is true. The learner studiously ignores the example of the skilled practitioner, and the skilled practitioner returns the compliment.

Make the learner a little more attentive without giving him any other new skills, and he's very likely to introduce many errors into his more frequent imitations. (Only luck could save his family if he tried to copy someone else's way of keeping a fire.) He's also likely to use the wrong version of the wrong tool on the wrong occasion, because the mere opportunity to imitate is no substitute for an explanation of what to do with it when and why. So in the case of any particular cultural behavior, error catastrophe would probably be its fate. Eventually—in fact, rather quickly—in the absence of any method of preventing it, the series of imitations of imitations would deteriorate into random nonsense, as they do in the children's game of telephone.

The errors that the learners introduced wouldn't have to be large. Each one might allow a version of the behavior that was *almost* as good as the version being copied. After a sufficiently long series of almost-as-good copies, they would still end up with something that was pretty bad. The same thing would happen to every cultural practice that was passed down in this way. The larger number of cultural behaviors in which these uncritically imitative chimpanzees were engaged would, over time, come to consist mostly of random, nonsensical, useless manipulations of objects; the inappropriate performance of behaviors that would be useful under different circumstances; and other pointless acts, including the sending of distorted signals in inappropriate situations.

That's the rosiest scenario, because I've considered only noise and not parasitism. In fact, among such uncritically imitative chimpanzees, the quality most likely to make a behavior transmissible would presumably be noticeability, not utility. Consequently, if we want to take the idea that culture can evolve seriously, we must suppose that most cultural behaviors in this sort of very chimpanzee-like but more imitative creature would be selected to be conspicuous and nothing else, that even behaviors harmful to the fitness of the individuals who engaged in them could easily become common. Remember that most possible behaviors are likely to be maladaptive, because there generally are more ways of failing than succeeding at any complex task. It's easy to imagine a proliferation of loud, exhausting, useless forms of display.

With such a burden of maladaptive and useless culture, the animals' fitness might be seriously impaired, which, I suppose, is why no such slightly less individualistic chimpanzees exist. For a typical chimpanzee in a typical population at a typical moment in history, emulation is a much safer way of gaining skills, because as an imitator she would be inhumanly credulous and careless and, in a population of other inhumanly credulous and careless imitators, would learn things that were mostly useless or dangerous. (The importance of fidelity has been emphasized, among other places, in Boyd and Richerson 1985; Dawkins 1976; Dennett 1995; and Sperber 1996.)

The lack of any error-correction mechanism might well make mutation pressure, not selection, the dominant force in the evolution of chimpanzee culture. In that case, in a population of uncritical imitators, an increased propensity to imitate would be, on average and for the typical individual, nothing but a greater propensity to engage in elaborately random behavior. In the African wilderness, this would probably be a rather slow and chancy form of suicide. It's hard enough to survive there even when your actions are exactly the right ones, so randomly engaging in pointless behavior and making that your way of life is likely to be fatal. It seems natural to us, as humans, to assume that culturally transmitted behavior would be better behavior, but it isn't obvious that this is a law of nature.

Given the very poor fidelity with which apes replicate observed behaviors in the absence of human teaching, it's hard to see how any cultural tradition in an ape society that was transmitted by imitation could possibly escape such a fate, could possibly escape "error catastrophe" (Eigen 1971, 1992:20; Eigen, McCaskill, and Schuster 1988) and a rapid descent into maladaptive randomness, which may explain why all apes almost always prefer to learn by emulation. Emulation poses less danger because only behaviors that immediately produce inherently desirable results will be adopted as a result of it.

For a second, cultural, system of inheritance to function well, it presumably would require a second system of error correction, analogous to the various DNA repair and error-correction mechanisms, including sexual recombination, that we now know to be indispensable parts of the first, biological, system. The reader will recall, from the discussion of Manfred Eigen's error-catastrophe model in chapter 4, that these mechanisms are needed to reduce the mutation rate so that it will be lower than the selective disadvantage of the slightly mutated versions, allowing natural selection, rather than mutation pressure, to become the dominant force in the organism's evolution. To begin to accumulate culture, our ancestors would also have had to reduce the "mutation" rate in learned cultural behaviors to a level that was lower than the relative selective disadvantage of the slightly corrupted version of the behavior that the mutation would produce. As I said before, this is a problem that any living thing would have to overcome in creating a new form of heritable information, just as gravity must always be overcome to achieve flight.

Apes other than humans show no sign of having evolved any adaptations for faithfully transmitting culture between generations that parallel our own—aside, that is, from simply *not* transmitting it. They can be trained, by assiduously error-correcting humans, to use signs or engage in other human cultural behaviors because they're extremely clever, but training them is difficult, and they make plenty of mistakes in the process of learning. These mistakes are corrected only because we're in the picture. Without human intervention, ape sign language idiolects would rapidly diverge, becoming mutually incomprehensible within a generation or two.

An ape very seldom corrects another ape's mistake. Even if she's cognitively capable of doing so, the desire seems to be largely missing. Consequently, the "mutation" rate, the rate at which unfaithful copies of a behavior are produced in any imaginable process of cultural transmission among chimpanzees or organisms like chimpanzees, would be quite high. Unless and until some exogenous accident or change in behavior propelled a group of apes over the high fidelity threshold needed to avoid error catastrophe, the direct imitation of complex behaviors would be a very unsafe strategy for individuals to pursue and would be selected against. The typical imitator in a population of unfaithful imitators would be copying an already corrupted and useless version of a behavior.

If this line of reasoning is correct, we should expect to find either very imitative apes in nature, with all sorts of specialized cognitive adaptations for improving the fidelity of cultural transmission, or apes who don't imitate much at all, who generally just aren't willing to. Either copying fidelity is high enough to get the population over the error-catastrophe threshold, in which case cultural evolution is possible and probably will already have been going on for millions of years, or it isn't, and cumulative cultural evolution can't happen, even though individual items of culture may be invented and conserved for a little while by the much safer strategy of emulation, or through audience effects on the process of ritualization.

It would be extremely surprising to find a group of apes with an evolved, stable ability to have a *little* culture that they weren't very good at passing on, with no specialized adaptations for transmitting it in a faithful manner. Unless the behavior was adaptively inconsequential or the method of cultural transmission contained an intrinsic form of error correction, evolution would quickly push any such population in one direction or the other. Either they would fairly quickly evolve an elaborate set of adaptations for error correction and faithful transmission, or they would stop imitating each other altogether, depending on how much selection there was for the maintenance of the capability.

And of course, what we actually do find in nature is exactly what we would expect. Humans imitate a lot, and humans correct one another's mistakes a lot. Chimpanzees and other apes don't imitate very much; they mostly emulate, and they very seldom correct one another's mistakes. This is not because they're too stupid to imitate—even a bird is smart enough to do that. It must be at least partly because the fidelity of their imitations would be too low to avoid error catastrophe, and make imitating a good idea for the typical individual in the typical population.

Even if fidelity magically became higher, it's hard to imagine a young chimpanzee (or even a young human child) being a particularly good judge of whether a complex imitable behavior is adaptive over the long term. Waving burning branches in the air may seem like an impressive display, but would unsupervised imitation by a child be worth the risk? In contrast, *emulation* is the reinvention of the skill by every individual who masters it. The process of reinventing it includes extensive trial and error, so an intrinsic error-correction mechanism is built into the process of transmission. Only ways of performing tasks that the individual already is innately motivated to do, such as eating termites, can be successfully emulated. Only repeated and direct rewards can stabilize the behavior. So it's unlikely that much maladaptive or simply random behavior will creep in through this channel. The problem is just that by human standards, this is an incredibly inefficient method of transmitting skills between generations.

By making the acquisition of each skill so difficult and time-consuming, evolution has drastically limited how many learned skills each chimpanzee can possess and has virtually ensured that skills will be lost at a rate that keeps the absolute number of distinct ones in circulation low. Safety has been achieved at the expense of complexity. If chimpanzees needed just a few more tools to get by, if their toolmaking culture needed to become a little more elaborate in order to cope with some selective challenge in a slightly different environment, they would have trouble responding, because they're already learning as much as they can, given their very cautious and antisocial way of doing it. In that situation, something better and faster though riskier than emulation would have to evolve—but it couldn't, unless a more efficient way of dealing with the problems of poor fidelity and maladaptive culture happened to evolve at the same time.

TEACHING, SHARED GOALS, AND RUNAWAY SEXUAL SELECTION

The link between the need for error correction and the distinctive human tendency to pursue shared goals is simply that it may be inherently easier for one of us to spot the errors of another—in particular, for an adult to spot the errors of a child—than it is for us to spot what we ourselves are doing wrong or for the child to correct his own mistake. It may be easier for a teacher, in the broadest possible sense of the word, to weed out inadequate versions of culturally transmitted behaviors, particularly inaccurate copies of the teacher's own skillful behavior, than it would be for the learner to weed them out by herself.

Repairing DNA copying errors and other mutations in cells often requires using a second copy of the original sequence as a template for correction. (The sequence used may not be a direct copy; it might also be a complementary sequence like the one on the damaged DNA molecule's other strand.) Otherwise, how is the cell supposed to know which changes count as "repair" and which as additional mutations? If error correction in the transmission of human culture is anything like that, if the repairing of mutations and the filtering out of maladaptive culture are typically pursued, in the transmission of human culture between generations, in a manner that involves coordinated activities by two or more humans to allow reference back to an uncorrupted copy of the cultural trait or behavior, then the need to filter out mutated and maladaptive versions of skills would adequately explain our propensity to coordinate our activities around shared purposes and intentions.

Intensified cooperative activity might have started for another reason, perhaps to exploit some new source of food available only through collaborative tool use, as Sterelny (2012) suggested. (Behind elaborate forms of sociality in animals, there often is an opportunity to obtain more food by cooperating, by storing nectar as honey, for example, as well as cooperating in the rearing of young.) As the indispensable prerequisite for a more elaborate form of culture, however, cooperative activity would have been preserved by the need to keep any culture whose evolution it might enable going. The constant and continuing need for mutation repair and the filtering out of pointless or parasitic behaviors would keep it in existence. From that point on, only relatively cooperative modes of life were open to us.

In Maynard Smith and Szathmáry's (1998) terms, the cooperative filtration of culture is an adaptation that has "contingent irreversibility," the phrase they use to describe ways of behaving that become hard to abandon once started, often for reasons that have little to do with the reason they began. Once a new form of cooperation has appeared, parts of the organisms involved that were once essential may become redundant and be lost, or a new way of life may be adopted that is possible only if cooperation is maintained. Once the cooperative filtration of human culture began, and a set of culturally transmitted skills became essential to survival, there would have been no going back, no way of becoming a fully competent adult, or producing fully competent children, without engaging in it.

Female bonobos already cooperate to deal with male aggression (Fruth and Hohmann 2002; Hare, Wobber, and Wrangham 2012; Hohmann and Fruth 2003a, 2003b). Evolving the ability in adults to cooperate with the group's younger members in the high-fidelity transmission of culture between generations doesn't seem much more difficult, given the right ecological circumstances, especially if, as Sterelny proposed, the group needed to cooperate in food gathering, food processing, or some toolmaking activity that the younger members could be taught to join.

In chapter 2, I suggested that developments in game theory occurring after David Lewis wrote *Convention* ([1969] 2002) give us reason to worry that he may have been too optimistic about our ability to spontaneously converge on the best forms of coordination in the absence of any policing. I pointed out that Ken Binmore and Larry Samuelson (2006) demonstrated that if nature is allowed to take its course—if paying attention to environmental cues is costly, if how much attention the players of a coordination game pay to which sorts of environmental cues is left up to the individual players, and if no sanctions are applied to players who don't pay enough attention or attend to the wrong thing—then the players will not pay the optimal amount of attention to the right cues. Instead of arriving at the optimal, "payoff-dominant" level of attentiveness, players will evolve toward paying a smaller, "risk-dominant" amount of attention to the cues needed for coordination, even though

they may often miss cues that would allow them to coordinate successfully in particular situations.

Again, the basic reason is that when a player who is paying close
attention to many cues is paired with a player who is paying slightly less
attention, it is the attentiveness of the inattentive player that determines
the probability that coordination will be achieved. The inattentive player
at least is spared the costs associated with attending carefully, but the
attentive player gains the exact same benefit from their interaction but
pays a higher cost for monitoring the environment. Since slightly less
attentive players do better against slightly more attentive ones than the
more attentive ones do against them, they can afford to do slightly worse
in their interactions with their own kind than more attentive players
do in interactions with *their* own kind, especially when the fraction of
inattentive players in the population is low. Thus a population of more
attentive players can be invaded very easily by slightly less attentive
players. Through a series of small steps like these, one after the other,
the population will inevitably move away from the optimal, "payoff-
dominant" equilibrium of paying close attention to the cues needed
for coordination to a more inattentive, "risk-dominant" equilibrium in
which everyone is worse off.

I suggested that the situation of parties who need to engage in complex forms of coordination with one another might be improved by
an attentional equivalent of the highway patrol, to weed out these less
attentive individuals or at least encourage them to be more attentive. I
then pointed out that modern human parents, teachers, and others with
whom we converse or coordinate do sometimes sanction or reproach
us if we fail to pay attention to the things we're supposed to be paying
attention to.

In Sterelny's scenario of intensified cooperation in food gathering
or food processing, the amount of attention that some of the cooperating parties were paying to particular cues for coordination could easily
come to be important to the other parties in a way we don't really see
in chimpanzees.

The direct eye contact we sometimes see in sexual interactions
between bonobos occurs in the context of a natural opportunity for
one bonobo to take an interest in what another bonobo is attending
to, and perhaps even to the fact that the other individual is attending

to what he himself is attending to. This relatively intense attention-to-attention—perhaps even involving attention-to-attention-to-attention, a simple form of recursive mind reading—seems to be something of which this descendant of the common ancestor of chimpanzees and humans is already capable. What an animal can do in one behavioral context, she and her close relatives may also be able to do in another.

Chimpanzees don't learn to employ all the gestures that members of their own group use partly because no one is forcing them to do so. In the absence of an obligation to learn or employ a whole language, L, the available communicative conventions are sparsely observed, like the convention of going to lunch at noon, partly because they're nonbinding. They're private arrangements of convenience between particular individuals, not clauses in some conventional extension of the social contract. In contrast, our present-day binding convention of telling the truth in a particular public language includes an obligation to learn all its common words, not only that two lanterns mean arrival by sea, but also that one lantern means that the British are coming by land. (See Putnam's [1975b:248–49] discussion of what we all are obliged to know about tigers.)

The difference between a nonbinding and a binding convention, I argued in chapter 2, is in whether or not clauses 4 and 5 of Lewis's ([1969] 2002) definition—that we prefer other individuals to conform to the convention, provided everyone else is conforming, but if we had some other convention, we'd prefer them to conform to that—have any teeth. Apparently, one thing needed to progress from the sort of sparsely observed, nonbinding Skyrmsian communicative conventions that chimpanzees have to the binding and universally observed conventions of a modern human language is a preference that others conform to the same communicative conventions and perceptions regarding what's salient as we do, plus a willingness to do something about it. The conventions used for communication must stop being arrangements of convenience worked out between pairs of individuals and begin deriving their validity from the preference of all the members of the community that everyone should observe the same conventions, that everyone should go out to meet the British when the signal is given. These conventions must become associated with a social contract, and their users must begin weeding out inadequate attempts to comply with that contract.

From this beginning, it's a natural step to a shared preference that others see the same analogies between objects and events as salient and "natural," in Lewis's ([1969] 2002:37–38) sense. It's a natural step to preferring that others (particularly children) see firewood as firewood or hammers as hammers or digging sticks as digging sticks or spears as spears or corms as corms, and thus to binding, universally observed conventions that there *are* such things—firewood, hammers, digging sticks, spears, corms—and that they have specific names. Inherent in the very idea that we must hunt the stag together, and not separately hunt rabbits, is a presumption that we all know what a stag is and what a rabbit is and that we all would assign the same objects to each category.

Gergely Csibra and György Gergely (2006, 2009, 2011; Gergely and Csibra 2006) have argued persuasively that modern humans possess a set of psychological adaptations for teaching, in the broadest possible sense of the word, including very informal and casual forms of correction or demonstration, and for learning from those who would teach us. I will describe a few of these ideas in a later chapter. For now, the point is that any sort of deliberate teaching amounts to a form of domestication of culture, because the teacher often chooses what to pass on and doesn't pass on just any nonsense, and the pupil often chooses which teacher to attend to and doesn't learn from just anyone, skilled or not. Both parties have an effective veto over the reproduction of the item of culture. On the basis of human preferences and perceptions, the teacher culls unattractive or imperfect versions of cultural behaviors with an expert eye and promotes the reproduction of preferred behaviors, selectively rewarding or praising performances that are of unusually high quality. Rewarding those who pay attention to the right things and dealing with those who don't seems like a method of filtration that might be adequate to avert error catastrophe, to exclude at least some maladaptive cultural behaviors, and to create a universally shared set of binding conventions. Interactions that have the effect, on the individual level, of equipping each of us with an acceptable set of culturally transmitted skills and shared classifications may also, on the population level, keep noise and parasitism from overwhelming the system. Without such an adaptation, the accumulation of substantial amounts of uncorrupted, useful, universally shared culture may well be impossible.

If this is true, it would help explain both the nature and the possibility of the major transition that took us from chimpanzees to humans. At its heart was a symbiogenic fusion between humans and our culture that was made possible by this newly evolved cooperative culture-filtration equipment, perhaps loosely based on the simpler intention-attributing machinery found in chimpanzees but complicated by a capacity for recursion, by the ability to think about what we know that the learner doesn't know and doesn't know we know, and attached to a completely new set of motives, a willingness to manage the attention of others in a way that's beneficial to them and not just ourselves.

A generic form of altruism doesn't seem to be exactly what was needed to accomplish this transition. It seems more accurate to say, as Sterelny seems to want to, that we had to become more *helpful*. The important innovation wasn't a willingness to engage in behaviors that would reduce our own fitness; it was a willingness to engage in behaviors that would increase the fitness of others, possibly while increasing our own fitness or the fitness of our close relatives at the same time. In this story, there's no need to invoke group selection. As Tomasello (2008) argued, we simply needed to become willing to point things out to others and to have them point out things to us.

If we take seriously the idea that chimpanzees are on the far side of a major transition from us, then the problems that we humans solved to fuse into the evolving cognitive network we call a culture are precisely the problems that chimpanzees must have dealt with by avoiding them. We humans are helpful but critical, enjoy working together on common projects, and like to point things out to one another. In contrast, chimpanzees aren't very helpful, aren't critical at all, don't seem to enjoy working together, and are largely indifferent to attempts to point anything out. It seems plausible that by avoiding it entirely, they've solved the same problem we humans solve by helping one another and working together. This problem, I believe, is maintaining a culture across multiple generations without its being overwhelmed by noise and parasitism. The ability of one helpful human to act as a filter for the cultural inheritance of another, often more helpless human and, while doing this, unselfishly manage the attention of that other human strikes me as the key to overcoming it.

If that's the right way to think about the difference between humans and chimpanzees, then when we look back at humans again, what we should expect to see is a new set of collaborative machinery for improving the fidelity with which culture is transmitted, which also functions as, or comes packaged with, a system for excluding useless, pointless, maladaptive items, and discouraging experiments with noncooperative behavior. One thing that must be required is a clean template, an uncorrupted version of the behavior.

Over the course of a long life, a human being is capable of metamorphosing into a creature very well equipped to serve as a template and filter for her culture, because we continue learning and refining our inherited skills throughout our lives. The very long postsexual phase in human life history, missing in other apes, looks quite suspicious once we realize that such benign and wise and experienced creatures may be needed as part of our adaptation for cumulative culture. (For a summary of their very interesting theories about the importance of grandmothers in human evolution, see Hawkes and Blurton Jones 2005).

I argued earlier that the idea of domestication made sense as a description of our activities in the present, but now I believe I've also demonstrated that there is a place for it in our theory of human evolution. Domesticating, culling, and cultivating cultural behaviors would work as a way of preventing them from deteriorating into noise, just as this works in the very small populations of pure-bred dogs, and might even allow us to exclude some of the more obviously parasitic ones. But you would need to be clever to do it well, and a lifetime of experience would help. In fact, there's no upper limit to how clever you might benefit from being, because the quality of your culling can always be improved. You could show off your cleverness, if you were clever enough, by conspicuously consuming carefully cultivated culture (singing well, having a nice hairstyle or an elegant lip plate, saying clever things, telling good stories), and no doubt another creature like yourself would find that enthralling. Who could be a better parent or teacher for a human child than someone who clearly is already very skilled and accomplished, already the result of good parenting or teaching.

Darwin ([1871] 2004) thought that much of human nature could be explained by sexual selection, as we can see from the full title of his book on human evolution, usually referred to as *Descent of Man*. Its actual name is *The Descent of Man and Selection in Relation to Sex*. The first part of the full title makes the case that humans are related to the other primates. The second half gives an account of sexual selection in nature in general, and then argues that it must have been important to human evolution. This may have been his answer to Alfred Russel Wallace's argument that human intelligence exceeds what's needed for the daily lives of uncivilized people, which apparently led Wallace to doubt that it could have evolved by natural processes.

The hypothesis about culture, in general, that I am introducing here seems to give us a new way to agree with Darwin. As individuals, we reveal our own prowess as culture-farmers in a way that's hard to fake, through the beauty of the skills and attitudes we personally cultivate. We learn to play the guitar, paint our faces, and dress "nicely," to speak well and be polite. If all this arduous cultivation of culture is part of human nature, it seems impossible that humans would be blind to the differences among human individuals that the process produces. Instead, we might expect sexual selection for the ability to cultivate culture. We might expect to be strongly attracted to the singer of the spellbinding song, the clever talker, or the well-dressed person. The ill-mannered, ignorant, incompetent, boorish, poorly dressed, unhelpful, and completely uncultured person, however, the unimpressive, unfunny bumpkin who has no good stories, games, or tricks, has a harder time winning our hearts. (This seems to be a central theme of *The Tale of Genji*.)

It isn't that each song, adornment, or skill necessarily enhances fitness in some intrinsic way. Like chimpanzees' leaf clipping and ritualized invitations to play, most such actions probably are adaptively inconsequential. Their display tells us about a person's capacity to learn things, to pick up new skills and items of culture, and therefore it also may tell us about that person's capacity to teach. Like the nightingale's many songs, our own songs or stories or dances or skills are partly advertisements of our quality, inherently difficult things that we have been able to master in a way that has a pleasing effect on the senses.

Our ability to play music, dance, compose poetry, tell stories, decorate tools, and engage in other apparently useless kinds of culturally transmitted behaviors certainly seems to go far beyond any obvious immediate utility for hunting and gathering, or even warfare. These talents for incredibly elaborate forms of adaptively inconsequential behavior, incredibly elaborate forms of play, might be at least partly explained by the evolutionary arms race (Dawkins and Krebs 1979) that would certainly ensue if the ability to cultivate culture in a better way than rivals could became attractive in a mate.

This scenario is a bit like Ronald Fisher's (1930) "sexy sons" hypothesis about the role of display in sexual selection more generally. Individuals of the opposite sex may begin to prefer some visible trait that gives a small selective advantage, because the preference confers fitness benefits on their offspring. The preference will become more common, since the preferred trait is adaptive. That preference itself will begin to confer an additional advantage on those who carry the trait, especially those with slightly exaggerated versions: "Fanciers always wish each character to be somewhat increased; they do not admire a medium standard; they certainly do not desire any great and abrupt change in the character of their breeds; they admire solely what they are accustomed to behold, but they ardently desire to see each characteristic feature a little more developed" (Darwin [1871] 2004:651). As the trait spreads, the genes for the preference will spread along with it, because the individuals who carry it will be increasingly the result of matings between carriers of the trait and those who prefer it in a mate. The more individuals who carry the new preference, the greater the advantage of exaggerated versions of the trait will be, so there's a positive-feedback process that's likely simultaneously to make both more common. Fisher (1930) contended that if nothing gets in the way, there is in any situation "in which sexual selection is capable of conferring a great reproductive advantage, the potential of a runaway process which, however small the beginnings from which it arose, must, unless checked, produce great effects, and in the latter stages, with great rapidity" (137).

Fisher was thinking about bright plumage in birds, but in the human case, culturally transmitted skills and attitudes could easily have played a similar role. This sort of runaway selection process might help explain the great speed with which a very exaggerated version of the human propensity for cultivating culture evolved over the past few hundred thousand years.

In that case, in making his three-way analogy in *Descent of Man* among the evolution of personal adornments in human cultures, the evolution of sexual displays in birds and other animals, and the evolution of ornamental plants and animals under domestication, Darwin wouldn't have merely been pointing out an interesting set of coincidences.

He may have noticed a three-cornered similarity between the two sets of human cultural behaviors and the displays of the birds that could tell us something about the nature of human culture itself. Apparently, culture must be cultivated in order to accumulate, so perhaps humans, in order to become human, had to begin to find the ability to cultivate culture attractive in a mate.

If a system like this is really necessary, if cumulative culture is useless and even dangerous until and unless it's domesticated, why has nobody pointed this out before now?

In the final analysis, the question of whether domestication is really necessary is a question about the mechanics of cultural evolution: Are noise and maladaptive culture really the problems that Maynard Smith and Szathmáry's (1998) theoretical framework suggest they should be? Mere verbal argument doesn't seem to be enough to establish that. Eigen's (1971, 1992:20; Eigen, McCaskill, and Schuster 1988) model of error catastrophe or the very general considerations about the evolution of parasites and pathogens that I have mentioned come from subjects far away from the study of human cultural evolution, so some readers might think it is reckless to draw grand conclusions about human nature from such far-fetched analogies, however suggestive they may be.

Fortunately, however, the people who study cultural evolution already have thought about the question of what sort of cognitive capabilities are needed for the evolution of a human cumulative culture, and they have more than mere verbal arguments to support their conclusions. Despite their slightly different emphasis, some of the models contemplated in that literature seem directly relevant to this discussion. The proponents of these models, too, are worried about loss, corruption, and bad behavior. They, too, appear to have concluded that culling of some kind is indispensably necessary. Nobody but Daniel Dennett is explicitly talking, yet, about "domestication" as a possible solution, but if you look at the models without preconceptions, some of them do seem to be moving in that direction.

6

The Problem
of Maladaptive
Culture

AN EVOLUTIONARY ARMS RACE?

The idea that cultural evolution might not always increase the Darwinian fitness of the organisms whose culture is evolving is fairly old. Luigi Luca Cavalli-Sforza and Marcus Feldman (1981) were discussing it more than two decades ago.

Why should we worry about this? The basic problem is that cultural behaviors are more numerous than human individuals and seem capable of reproducing more quickly than humans can. Our opportunities to have children are considerably fewer than our opportunities to imitate. With a larger population and more opportunities to reproduce, they should evolve more quickly than we do. If we look at the past five centuries of the world's history, our cultures certainly appear to be changing much more quickly than our genetic endowments are. This puts us in a very poor position to compete with culture, head to head, in an evolutionary arms race (Dawkins 1982; Dawkins and Krebs 1979; Fisher 1930). It's hard to win a race against something that moves much faster than we do. If it can compete with us at all, our culture should be able to out-evolve us in the same way an influenza virus does. That would make most human culture actively parasitic, detrimental to our health or hope of reproducing. From a Darwinian point of view, the best thing for us to do in that case would be to become as completely unresponsive to opportunities to acquire culture as chimpanzees are. Why hasn't our culture defeated us in this evolutionary arms race, if it hasn't?

Are there any other examples in nature of coevolutionary arms races that ought to have been easy for one side to win, for the same general reason, but that also, surprisingly, have ended up in stable mutualisms? One example that might fit this description is the relationship between termites and the microorganisms that live in their gut and help them digest cellulose. Because they're small and numerous and reproduce rapidly, the microorganisms easily should have won an evolutionary arms race with the termites and become pathogens, but they've been living together in a stable mutualism for hundreds of millions of years.

Why? Questions about exactly why things happen are always hard to answer in evolutionary biology, but in this case, one behavior looks very suspicious. Every time a termite molts, it loses all its intestinal microorganisms. Each time, its supply must be replenished by a process called *trophallaxis*, which basically is an exchange of stomach contents between two termites. Presumably, this process allows the termite to only engage in horizontal gene transfer—the transfer of the genomes of the symbiotic microbes—with obviously healthy partners. The impossible task of examining the genome of the transferred microorganisms is replaced with the simpler task of examining the source termite for any obvious signs of ill health, as it is in the parallel case of sexual selection.

Microorganisms that produce noticeably sick termites, termites that don't smell right, probably aren't passed on. The specialized microorganisms in a termite's gut can't reproduce in any other way; they are passed on either by trophallaxis or not at all. This mutual inspection is therefore a form of culling, the exercise of a veto over the microbes' reproduction that for hundreds of millions of years has allowed termites to routinely exchange microbial symbionts in a way that otherwise would be very likely to spread disease.

As with sexual selection, in termite trophallaxis the animal's brain, such as it is, must decide whether a partner is strong enough to be a

good risk as a source of (in this case, microbial) genes. The brain is acting as a filter in the flow of genes. If we told a parallel story about how
humans overcome the risk of maladaptive culture, it presumably would
have to include something like this, some use of the human brain to
gate the flow of culture and weed out dangerous items, partly by avoiding impaired sources. This, of course, is one of the motives for thinking
in terms of domestication, since its essence—the control of one replicator's reproduction by another—is the imposition of just such a veto,
just such a gate. I don't think it's unreasonable to say that termites have
domesticated their microbial symbiotes, in the broader of the two senses
of the word *domesticate* that I explained in chapter 1, since they can't
survive without the termites and the termites can't live without them.
This means that the microbes themselves can't survive without being
repeatedly culled by the termites in this way.

Colonies of social insects have, effectively, a shared stomach (honey
is regurgitated nectar, and ants exchange trophallactic "kisses"), while
colonies of humans have a shared mind. We routinely disgorge some of
the contents of our own minds into the minds of others. That is what
"culture" is. Perhaps establishing a secure high-bandwidth connection
between two or more minds (of the sort depicted in the painting used
as the frontispiece for this book) also requires the existence of security
checks and border protocols at the point of exchange, just as sex or
trophallaxis does.

This is where a search for parallel cases in the biology of other highly
social organisms might lead us. Cavalli-Sforza and his colleagues' proposed solution to the problem of maladaptive culture (Hewlett and
Cavalli-Sforza 1986) was rather different. It invoked a feature of our
ancestors' social lives that's common in nature but, in other animals,
isn't associated with the solution to this sort of problem. They thought
that maladaptive culture might not have been a problem for our ancestors because their societies were small. They argued that much of our
culture must have been transmitted vertically in the small societies in
which humans evolved, that in more traditional societies we must have
received most of our skills and values, along with our genes, from our
parents. As long as genes and culture are acquired from more or less
the same people, they suggested, both will necessarily have to have been
adaptive in previous generations.

The consequence that people tend to draw from this argument is that the maladaptive aspects of culture should usually be found in more modern societies, in which Darwinian selection and the transmission of cultural practices have come apart, because our cultural inheritance is no longer obtained mainly from parents and close relatives. In a modern, agricultural human society, institutions like universities and monasteries can promote the imitation of practices that don't maximize the production of offspring. Our less civilized ancestors, living in small groups in which personal survival and the survival of the group's culture were closely linked, supposedly didn't have the luxury of engaging in maladaptive cultural practices, so the problem didn't arise.

It now seems doubtful that vertical transmission is as important as Barry Hewlitt and Cavalli-Sforza originally supposed. Even though their hypothesis may still have adherents, many people who study cultural evolution are now skeptical of its ability to solve the overall problem of maladaptive culture, mostly because it's difficult to interpret the evidence as saying that most transmission of cultural practices is vertical in any human society (Aunger 2000; Boyd and Richerson 1985:53–55; Chen, Cavalli-Sforza, and Feldman 1982; Lancy 1996; Plomin et al. 2001; Richerson and Boyd 2004:148–90).

I certainly haven't been able to persuade myself that it solves the problems I described in the last two chapters. I can't see how vertical transmission would solve the problem of noise, as opposed to parasitism, at all. Manfred Eigen's (1971, 1992:20; Eigen, McCaskill, and Schuster 1988) model of error catastrophe already assumes that transmission is vertical. But as far as I can tell, it wouldn't really solve the problem of parasitism either, because HIV can be transmitted vertically, from a parent to a child. Intracellular parasites of germ-line cells such as *Wolbachia* can be transmitted *only* vertically. The existence of any kind of intragenomic conflict depends on the fact that vertical transmission alone isn't adequate protection against parasitism. What matters in these cases isn't whether transmission is vertical, but whether it's perfectly Mendelian, a much more delicate and difficult-to-achieve condition. The human way of doing things doesn't seem to involve any arrangements for making cultural inheritance Mendelian, making sure that we receive exactly half our culture from our father and half from our mother, in every generation.

The hypothesis also is inconsistent with history. If vertical transmission was once the only thing keeping human culture from becoming maladaptive, and modern culture is therefore more likely to be maladaptive than hunter-gatherer culture, why has population growth been so much more rapid in modern times, when the hypothesis predicts that modern, horizontally transmitted culture should be lowering our fitness, not enhancing it?

The theoretical problem that Cavalli-Sforza and Feldman pointed out is a real one: put a gene or another replicator in a position to be selfish, and it probably *will* become selfish. Our first guess about how the problem might be solved, however, hasn't become widely accepted. In my opinion, there now are more plausible models in the cultural evolution literature to address the problem of maladaptive culture and what humans do about it. Although there are far too many for me to discuss them all, or even the most important ones, I can offer readers a very small and unrepresentative sample from that rich and complex tradition, one that I hope will show how the discussion is progressing. The conclusion will be that the quantity and quality of culling matter a lot, that nothing like the cumulative culture we see all around us could exist without a fairly sophisticated filtering mechanism to clean out some of the most dangerous or futile of our behaviors and practices.

IMITATION AND ADAPTIVE LAG

Even if we assume that the problem of noisy transmission between generations, and the error catastrophe to which it apparently should lead, has been resolved—as it must have been at some point in human evolution in order to produce creatures like ourselves—Alan Rogers (1988) demonstrated that uncritical imitation by itself won't, over the long term, increase the individual fitness of imitators or the mean fitness of the groups to which they belong. If uncritical imitation works as it does in his deliberately oversimplified model, it still can spread because it gives an advantage to particular individuals. In the longer run, however, it does nothing at all to increase the numbers of the population in which it arose and gives them no advantage in competing with populations that lack it. In a finite population over the very long run, it seems as if it could easily lead to extinction.

Rogers assumed a changeable environment. This is a common assumption in models of cultural evolution, since without it, it's hard to start selection for any kind of individual learning, cultural or otherwise. In an environment that's completely stable, why not just do everything by instinct, as ants do? Later, I'll describe a model in which culture becomes its own environment, so that changes in culture drive more changes in culture, but since African environments were quite changeable during the period when humans evolved in them (Finlayson 2009; Potts 1997, 1998; Richerson and Boyd 2002), the assumption of an initial impetus from environmental instability isn't unreasonable. Indeed, the truly stable parts of Africa, the parts still covered by rain forest, as the whole continent once was long ago, are still inhabited by truly stable apes, by gorillas and chimpanzees and bonobos, so it makes sense to think of our own ancestors as specialists in marginal or variable environments, as the desert elephants in Namibia are today. (These elephants also rely on learned information to find water in years when it's scarce. Among elephants in general, a family group's success in having and raising calves may be correlated with the age of the oldest female in the group [Foley, Petorelli, and Foley 2008; McComb et al. 2011]. This coincides with a human-like life span and life history, with many individuals surviving well beyond reproductive age.)

In this changeable environment, Rogers put a population of mythical creatures, "snerdwumps," who, to learn to perform a resource acquisition activity like foraging or hunting or seeking water, possess two genetically inherited strategies they might pursue. Some snerdwumps are born with a tendency to learn from their own experiences, and others are born with a tendency to imitate another, randomly chosen member of the population. For example, some snerdwumps might be willing to try novel foods or look for new sources of water, while others eat whatever they see their "cultural parent" eating or look for water wherever their cultural parent used to look.

If learning by imitation isn't less costly than learning through personal experience (which may involve a risk of being poisoned or dying of thirst), no snerdwump will learn by imitation. But if it is less costly, then the first mutant snerdwump who starts imitating others will receive immediate fitness benefits and, all else being equal, should have more surviving offspring than her peers do.

As long as the fraction of the population who are imitators remains small, these benefits will continue to accrue, and that fraction will continue to increase. At some point, however, some of the imitators probably will choose as their cultural parent an individual who herself is an imitator, who learned her approach by imitating some other individual. This is fine as long as the environment remains stable, but if it has changed in the last generation, the second-hand strategy that the imitator of an imitator adopts already will be obsolete. The greater the fraction of imitators in the population is, the more likely this is to occur, especially because some of them already will be imitating imitators, or the imitators of imitators, themselves.

Many of the imitators, the number depending on what fraction of the population they make up, will therefore lag behind in adopting the appropriate new strategy after a change in the environment, will go on looking for water where there currently isn't any. This lag will impose a load on the imitators' mean fitness, dragging it down toward the mean fitness of those who learn by personal experience. Eventually (without the mean fitness of those who learn by experience having changed in any way) the fitness of those who use the two different strategies will become equal, and the population will once again be at equilibrium, with the same mean fitness as it had before but now composed of a mixture of imitators, who learn cheaply but often lag behind, and experimenters, who have a more costly learning process but are more likely to be pursuing strategies appropriate to the current environment. That fraction will depend on how costly experimental learning is, how much cheaper imitation is, and how often the environment changes.

While the mean fitness of both strategies is the same over the long term, it can be very different in the short term, which means that in a world of finite populations, the innovators can easily become extinct at any time. If there's an unusually long spell of environmental stability, they may not survive it, because they'll be wasting effort, in every generation, to acquire skills that they could obtain more easily by imitating their neighbors. The water is always in the same place—why continue to waste effort on dangerous searches for new sources?

The innovators could get along fine without the imitators, but if all the innovators die out as a result of an unusually long period of relative stability, then the imitators will be in trouble, since they will never again be able to keep up with the changing environment. When conditions become more arid again, they'll have no way to respond. So over long periods of time, if the variability of the environment itself is fairly inconstant, finite populations of snerdwumps should occasionally become completely dominated by imitators and then become extinct.

Even if the cost of learning by imitation is much lower than the cost of learning by experimentation, and even if environmental changes are quite infrequent, the only consequence will be more imitators and a correspondingly greater lag between environmental changes and the widespread adoption of new strategies, not an improvement in the fitness of either imitators in particular or the population as a whole. Much of the time many snerdwumps will be doing the wrong thing in such a population, in a way that completely cancels out the benefits of imitation as a strategy.

The point of this argument isn't to show that cultural learning is impossible or pointless; it is to show that a successful system of cultural learning has requirements beyond the mere ability to uncritically imitate others, even if some members of the population are capable of creating useful inventions for the others to imitate. Rogers contended that his model could give us insight into what else was needed and shed some light on how adaptive we can expect cultural traditions to be at equilibrium. If experimental learning is quite costly relative to learning by imitation, then cultural traditions are unlikely to be very adaptive. But if it's cheap, or if the environment changes frequently, the competition from learning by trial and error should help keep any culture that survives up-to-date and fitness enhancing.

Whatever these additional requirements for adaptive cultural learning are, perhaps chimpanzees simply don't meet them. In that case, they aren't just stunted, handicapped humans with only a rudimentary ability to imitate. Instead, they're creatures for whom imitation would be a very bad idea, rugged individualists who innately "know better" than to copy their neighbors' possibly maladaptive or obsolete behavior. If that's correct, chimpanzees' rather rudimentary signaling skills don't show that they're simply not as bright as parrots; rather, they're the consequence of millions of years of selection *against* social learning, against excessively imitative or communicative chimpanzees. The question still is what the crucial differences could have been that set our own ancestors on such a different evolutionary path.

CRITICAL LEARNING AND ADAPTIVE FILTRATION

A nice, simple, revealing model of what the fundamental difference between snerdwumps and humans might be comes from Magnus Enquist, Kimmo Erikson, and Stefano Ghirlanda (2007). It's a fairly straightforward attempt to rescue Rogers's (1988) model. In their paper, the model is rescued by modifying it very slightly. Rather than endlessly persisting in blindly imitating the behaviors of some other individual whether or not they produced good results, what if cultural learners, after imitating an observed behavior, evaluated the results and abandoned those imitative behaviors that worked out badly when they were tried, switching back to experimental learning in those cases?

It may be easier to tell whether a technique is producing the desired results than it would be to invent it in the first place. ("I'm just not getting any termites! To heck with this!" The next time you feel frustrated, reflect on humanity's enormous debt to that emotion, and you will see that the impulse to curse and kick things is a noble one indeed.) As Enquist, Erikson, and Ghirlanda suggest, the problem with Rogers's model is that the only source of adaptiveness in the behaviors adopted is the original trial-and-error learning process by which they were created. But aren't the imitators members of the same species as the inventors, and don't they also have the capacity to observe outcomes and be reinforced or discouraged by them?

This apparently small change does indeed seem to rescue the model, leading to a population with a higher average fitness in the long run. If the initial process of inventing techniques by trial-and-error learning is made fallible, so that the techniques invented aren't always adaptive in the first place, these conclusions are further reinforced. To me, this seems to suggest that human culture is essentially, innately, and inextricably domesticated, because its very ability to be adaptive at all appears to depend on the existence of a distributed activity of deliberately weeding out unsatisfactory items. Only if we periodically become frustrated with particular items of inherited culture and discard them, stop cultivating them, and cull them—as our ancestors would have culled a two-headed calf or a blind sheepdog or a broken-down colt—can we have a human kind of culture at all.

Of course, other models—for example, preferentially imitating conspicuously successful individuals (Boyd and Richerson 1985:241–79)—might produce similar results, but I suspect that they all would require humans to make somewhat intelligent and rational choices among the available alternatives, thereby influencing which cultural behaviors would be passed on to the next generation. The one indispensable element is locally rational choice, myopic optimization, trying to choose the best version of the behavior to imitate, persist in, or encourage and to avoid the worse versions.

Enquist, Erikson, and Ghirlanda's modification of Rogers's model works (at least for individual resource-extraction skills, such as fishing for termites; it isn't clear how it could deal with collective-action traps, like unsafe mines), but it isn't the only way to arrive at a model of cultural evolution that works. Another paper from Enquist and Ghirlanda (2007) offers a more general argument about what is required.

An obvious difference between humans and chimpanzees is the *amount* of culture maintained in our societies. Despite continually inventing new ways of doing things as they seek to emulate those they envy, chimpanzees don't end up with very much culture. Human populations maintain many more distinct items of culture. What determines these amounts, and what makes them so different from one another? Presumably what we need to answer this question is a story about births and deaths, about new items being added and old ones being lost, and a balance between these processes determining the final equilibrium. Many people have made models of this general kind—Joseph Henrich's (2004) model of culture loss in Tasmania, after the small population there was cut off by rising sea levels at the end of the last ice age is a fascinating example. But do we need some kind of intelligent filtration, or culling, as well? Enquist and Ghirlanda constructed a rather simple model in response to this question and arrived at a strong and yet intuitively satisfying set of conclusions.

The authors were interested in modeling the dynamics of a pool of cultural behaviors. Let's suppose that new ones are invented at some rate γ per generation. Of these new "actions or rules for action, ideas, values, and artifacts," which they lump together as "cultural traits," a fraction q is adaptive, and a fraction $1 - q$ is maladaptive.

In every generation, a fraction λ of the entire existing pool of cultural traits is lost, not transmitted to the next generation at all. When λ is 0, every trait is successfully passed on. When λ is 1, no trait succeeds in being passed on.

In addition to being lost, an originally adaptive piece of culture could become maladaptive. The authors call this "corruption" and model it as occurring to a fraction θ of the existing pool of adaptive behaviors in each generation. Various things could cause this. The case they contemplate is environmental change. Behaviors that were adaptive under one set of circumstances may cease to be adaptive when the external circumstances change.

Rousseau's parable about the stag hunt, carried a little farther forward in time, serves as a perfect illustration of this problem. We can catch a stag if we all cooperate, but if one of us goes off chasing a rabbit for his own stew pot, the stag hunt will fail. How can each be assured that all will do their duty and that the stag will be caught, making it safe for them not to go chasing after rabbits themselves? I've already suggested that it's a lot easier to be confident that they will if something like clauses 4 and 5 in David Lewis's ([1969] 2002) definition of conventions are operating and have teeth, if people prefer that others don't run off after the rabbit and are prepared to punish those who do. In Enquist and Ghirlanda's paper, at least as I read it, this 260-year-old philosophical fable finally has an appropriate sequel.

After many seasons of cooperation, however we achieved it, the stag hunt has been so successful that we've killed all the deer in the valley, or the climate has changed in a way that's drastically reduced their population. If we don't go back to individually hunting rabbits, unsatisfactory as that is, we all will starve. But whatever cultural solution we came up with to get over the initial hurdle of mutual mistrust and create solidarity in the hunt still exists, preventing the rabbit hunting now needed for survival. There's a solemn compact to hunt deer together and never to betray the clan by hunting rabbits, or a ceremonial Deer King who leads the stag hunt, or a dietary taboo against eating rabbit, or a common idea that real men hunt deer, or something like that. A compact set up to hold together in the face of temptation might also hold up inconveniently well in the face of changing ecological circumstances. To survive as users of inheritable culture, we need a way of *abandoning* obsolete cultural achievements like our formula for ensuring cooperation in the stag hunt, whatever it is, in addition to ways of adopting these cultural achievements and passing them on to future generations. Remember the example of the unsafe mine. As Lewis ([1969] 2002:68–76) pointed out, conventions can easily become traps.

Of course, there are other kinds of "corruption" besides this sort of adaptive lag. The term also should probably be thought of as including mere deterioration into nonsense as a result of errors in transmission, as well as incremental evolution into an actively parasitic behavior. Loss, which is Enquist and Ghirlanda's λ, is the actual abandonment of a practice; corruption, their θ, is the practice becoming useless or dangerous in some way, as the stag hunt has in my extension of Rousseau's story, but still remaining in circulation.

These simple assumptions result in the following dynamics for the expected rates of change of the number u and v of adaptive and maladaptive traits, respectively:

(1) $\dot{u} = -\lambda u - \theta u + q\gamma$

(2) $\dot{v} = -\lambda v + \theta u + (1 - q)\gamma$

The terms $-\lambda u$ and $-\lambda v$ represent the loss of culture due to outright failures of transmission; the terms $-\theta u$ and $+\theta u$ represent the transformation of adaptive culture into maladaptive culture by corruption; and the terms $q\gamma$ and $(1 - q)\gamma$ refer to the original creation or invention of adaptive and maladaptive items of culture. Again, according to the authors, these are "actions or rules for action, ideas, values, and artifacts."

In this model, we tend to end up with more maladaptive items than adaptive ones because there's only one source for adaptive traits, invention at rate $q\gamma$, but two ways of losing them—failure to transmit, λ, and "corruption," θ—and two sources of maladaptive traits—corruption at rate θ, and invention at rate $(1 - q)\gamma$—but only one way of getting rid of them, failure to transmit at rate λ. It's assumed that maladaptive behaviors aren't "corrupted" into adaptive ones at any appreciable rate because there usually are more ways of failing to achieve some desirable result than there are ways of achieving it. Consequently, there's a unidirectional flow of inherited adaptive behaviors into maladaptiveness, with no countervailing flow going in the opposite direction.

Since the probability of learning something maladaptive is greater in this population than the probability of learning anything adaptive, cultural learning will be selected against, given most of the parameters' possible values. The point at which the benefits of the population's adaptive cultural behaviors are just about to be overpowered by the costs associated with its maladaptive ones can be found by setting $u\infty - v\infty$, the number of adaptive traits at equilibrium minus the number of maladaptive ones, equal to zero. Of course, this all depends on the per-generation loss rate, λ, because if behaviors don't hang around very long, they won't have much time to become corrupted.

With $u\infty - v\infty = 0$, for various values of the other two parameters we can find the value of λ, the fraction of behaviors that fail to be transmitted in each generation, at which there begin to be too many maladaptive and corrupted behaviors piling up, and cultural learning itself is no longer adaptive. Even for parameter values that do admit the possibility of *some* adaptive imitation (for example, if the organisms are very good inventors of adaptive behaviors and seldom invent any maladaptive ones) because of "corruption," we reach zero adaptive value while still at very high rates of per-generation loss, so genetic evolution will inevitably produce a situation in which the half-life of particular cultural behaviors is quite short, a few generations at most. To have even a little culture without being destroyed by it, the creatures that this model represents need to be both very good inventors and quite inattentive and forgetful; they need to be rather poor or unwilling imitators who seldom manage to pass on what they've learned.

This seems like a fairly good model of a population of chimpanzees. In the gestural communication system and tool-use traditions discussed in chapter 5, we certainly have plenty of evidence of continual, rapid social forgetting, of signals or tools being invented but not passed on to the next generation—but it doesn't look like the way humans do things. What's missing?

The authors point out that the situation changes dramatically if we add a term, ψ, reflecting some tendency to filter out behaviors that aren't currently working, by means of the critical learning process just described or by any other method. Then the dynamics for the expected number u and v of adaptive and maladaptive traits become

$$\dot{u} = -\lambda u - \theta u + q\gamma \text{ (no change here)}$$

and

$$\dot{v} = -\lambda v + \theta u - \psi v + (1 - q)\gamma \text{ (which obviously is different)}$$

This is a much better outcome. It turns out (again, for the mathematical details, readers are encouraged to look up the original paper) that if $\psi > \theta/q$, if the fraction of the existing maladaptive behaviors that are filtered out in each generation is greater than the ratio of the corruption rate of adaptive behaviors in each generation to the fraction of newly invented behaviors that are adaptive, then there's no loss rate λ so low that culture is no longer adaptive, because maladaptive behaviors are no longer piling up faster than adaptive ones at any value of λ, however low it is.

Apparently these creatures, unlike chimpanzees, can evolve the sort of cumulative cultural inheritance that humans have. Even if this condition for unbounded growth isn't met, much longer average lifetimes for cultural traits are possible. This means there's a way for an evolving population to move incrementally in the direction of an unbounded accumulation, as ψ gradually moves higher under the resulting process of natural selection.

This very simple and intuitive model implies that there are two distinct regimes in which cumulative culture can exist. If ψ is high enough, unbounded accumulation is possible, but if it's a bit lower, culture will accumulate until so much maladaptive culture is in the mix that the advantage it originally conferred vanishes. This situation would presumably select for a somewhat higher ψ, a slightly better ability to weed out maladaptive cultural traits, which would allow the accumulation of even more culture, leading to the accumulation of even more maladaptive culture, selecting for an even higher ψ . . . until finally ψ rose so high that the population broke through into the other regime, the regime of unbounded accumulation.

Obviously, the regime in which modern humans exist is one of unbounded accumulation, but it's interesting to think about whether at earlier stages in the evolution of our particular kind of ψ, when it wasn't as effective as it is now, we were ever stuck in the other one, the regime in which our carrying capacity for culture was limited by the associated load of corrupted or maladaptive practices. There was, as I've mentioned, a long period of relative technological stability in humans' recent evolutionary past, the period when our ancestors were using the Acheulean technology, which existed for many hundreds of thousands of years at a time with little apparent improvement. The earliest Acheulean tools we've found are roughly 1.7 million years old (Roche et al. 2003), and they continued to be used, without much change, all over the world until a few hundred thousand years ago. It's tempting to suppose that this long period of relative technological stability (I want to say stagnation, but the users of these tools were very successful animals, and we'll be very lucky if we last as long as they did) is an example of the other possible evolutionary regime for cumulative culture. Perhaps these creatures, unlike modern humans, really did carry a heavy load of maladaptive culture, which limited the total amount of culture that their societies could accumulate.

That's a bit hard to imagine, but the mere possibility is intriguing. In this scenario, at any given moment, many of their cultural behaviors already would have lost any benefit they once had and become counterproductive, limiting the amount of culture they could carry. Consequently, there would have been ongoing selection for a higher ψ. At some point, our adaptations for filtering culture—which must be one of the functions of our expanded brain—finally became good enough to propel us over the border into the other regime, the regime of unlimited accumulation that we still are in today.

What have we learned from all these models? It's tempting to suppose that the ancestors of humans initially differed from chimpanzees by just being better imitators of new cultural behaviors, but apparently this modification wouldn't have been enough, because by itself it's not productive of a stable, useful culture. More imitative chimpanzees probably would soon be worse, not better, at being chimpanzees. Much of their behavioral repertoire could easily come to consist of obsolete strategies for seeking kinds of food that are no longer common (perhaps because of the spread of the strategy or because the environment had changed), or versions of techniques so corrupted in transmission that they have become worthless, or maladaptive but conspicuous and tempting behaviors—for example, elaborate forms of display, the chimpanzee equivalents of alcoholism and smoking cigarettes. A long enough series of small, cumulative changes in these maladaptive directions would have added up to make them much worse off. We must have dealt with this problem somehow to become what we are. But how?

How do we deal with a biological virus or any other pathogen or parasite? Large organisms like humans became possible only when they evolved immune systems. All that food in one place has to be defended in order to continue to exist for very long. Similarly, an elaborate modern human culture seems to require the existence of some sort of "cultural immunity" system for filtering out maladaptive or corrupted behaviors. If we accept the idea that cultural evolution is a kind of Darwinian evolution seriously, the requirement seems inescapable.

Initially, the difference between the ancestors of humans and chimpanzees might have simply been an increase in some kind of ψ, the evolution of some set of cognitive or behavioral traits that were capable of acting as a filter to weed out behaviors that were no longer working. Maybe our ancestors somehow became better than chimpanzees at ceasing to imitate behavior that was no longer adaptive or that had never been adaptive, at deciding to give up doing things, or resolving to discourage others from continuing to do things. The question then would be what form this superior ability to abandon bad ideas actually took.

A number of models have been proposed, one of which is Enquist, Erikson, and Ghirlanda's (2007) simple "critical learning" model. But there are others. At least for modern human societies, it's probably necessary to be a pluralist and to think about many distinct selective processes going on at the same time.

From a philosophical point of view, what interests me is the existence of a universe of possible models that, like Enquist, Erikson, and Ghirlanda's critical learning model, ultimately cash out ψ in the form of countless human *judgments*, countless deliberate abandonments of old methods or conventions that aren't working any more, have been garbled in transmission, or were never really good ideas in the first place. Some of these judgments must take place in the minds of single individuals, but in other cases, coordination among multiple individuals would be required to abandon a practice that had become conventional or compulsory in a group.

Language sometimes would have been useful in arranging this coordination, so it seems plausible that the use of language has played some sort of role in the elimination of other kinds of maladaptive culture. We can point out that something isn't working. We can have conversations about what isn't working and why. Questions like the very common "Well, what am I supposed to do *now*?" and "Why do things always have to be this way?" can be asked and, in some cases, may even be answered.

The formal models of cultural evolution I examined in this chapter show that a human or human-ancestral society would have needed a way of abandoning outdated or corrupted pieces of culture in order to have much culture. Lewis's analysis suggests that once we began to have conventions—say, a convention of collectively foraging for some kind of food that had stopped being abundant—they often couldn't have been easily abandoned by any one individual acting on his or her own. Once everyone starts driving on the right, it's hard for any one of us to break the habit. Once you have developed the ability to hunt stags collectively, by adopting a convention of forming mobs to punish those who abandon their posts to chase rabbits, and you have killed all the stags in the neighborhood, how do you all move back to hunting rabbits, and how do you *stop* forming counterproductive mobs to punish individuals for doing what now is the only reasonable thing to do?

Think again of the unsafe mine, where all the workers would be better off if none of them showed up for work, but nobody wants to be the odd man out. What they all need to do is gather in a room and have a conversation so they can explicitly adopt a new form of coordination. Somehow, a new convention, revolving around a new precedent as its focal point, must become established, displacing the old one: "No, that's just not working any more, now let's all hunt rabbits instead." "No, conditions in the mine are too unsafe, let's all go on strike." Finding a method of abandoning solutions to coordination problems that have stopped working out well for us all is itself a coordination problem, one whose solution is required to keep the solutions to all our *other* coordination problems from becoming inescapable traps.

As Lewis ([1969] 2002:83–88) pointed out, a particularly good method of coordinating around a new convention, a fast and reliable way of making a behavior obvious, is through explicit statement. Conventions don't *have* to be tacit. The tacit conventions of language can be used to make more explicit conventions, or even to make themselves more explicit. What Thomas Schelling (1966) was originally interested in was explicit bargaining. Tacit bargaining was supposed to be a simplified model for this far more complex process. Every human society has tacit conventions, including many of the conventions of its language, but they all seem to have explicit ones as well, customary laws or shared narratives.

Whether maladaptive or corrupted cultural practices are abandoned as the result of explicit deliberation or prohibition or in other ways, Enquist and Ghirlanda's model suggests that human judgments must be acting as a mechanism of selection whenever this happens, filtering out some practices while allowing others to continue to be copied. We should remember the authors' point that it's easier to tell that something isn't working than it is to come up with something that will work. The judges responsible for making these determinations don't need to be geniuses, they just need to be experienced enough to know when something doesn't seem to produce good results anymore.

In chapter 5, I argued that one possible evolutionary explanation for the distinctive human tendency to pursue shared goals is that it may be inherently easier for one of us to spot the errors of another, for a skillful individual to spot the errors of an unskillful imitator, or for an adult or an older child to spot the errors of a younger child, than it is for us to spot what we ourselves are doing wrong or for the child to correct his own mistake. As I pointed out then, repairing copying errors in cells often requires using a second copy of the original sequence as a template for the correction. If correcting errors in the transmission of human culture is anything like that, I argued, if repairing mutations and filtering out maladaptive culture are typically pursued in the transmission of human culture between generations in a manner that involves coordinated activity by two or more humans and involves reference to an uncorrupted copy, then the need for filtering out mutated and maladaptive versions of skills would be one explanation for our propensity to coordinate our activities around shared purposes and intentions. Intensified cooperative activity might have been started for another reason, but as the indispensable

prerequisite for a more elaborate form of culture, it would have been preserved by the need to keep that culture going, by the constant and continuing need to repair mutations and to filter out pointless or parasitic
behaviors. Now I've added arguing (for example, about whether to hunt
stags or rabbits or whether to go on strike) and doubting (for example,
that this is the most effective way of finding termites or that stag hunting
is still worth the trouble) to teaching and explanation or demonstration
as additional examples of this sort of filter in human affairs.

Obviously these culling processes are similar to the process by which
a farmer or rancher decides which of his animals or plants should be
allowed to pass on their genes to the next generation. In those cases,
too, the lack of any ordinary process of *natural* selection makes it seem
likely that deleterious mutations would accumulate in the absence of
human decisions. Yet unfettered natural selection would turn our dogs
back into wolves and our maize back into teosinte. We don't want that,
but we also don't want irreversible deterioration, so we exercise some
care about which animals or plants we allow to breed. This—the control
of one replicator's ability to reproduce by another—is, as I pointed out
in chapter 1, the crucial part of "domestication."

The same cognitive capacities that enable us to pick and choose animals and plants in order to find those that seem the best also must have
been operating for hundreds of thousands or millions of years, when
we chose which imitative or newly created behaviors to discourage or
allow in ourselves or other people. (Our modern habit of domesticating
animals and plants would then, as I suggested in chapter 1, be a fortuitous side effect of a much older adaptation.) A parent or teacher or an
older child who discourages only a few behaviors a week is still culling
as actively as any farmer could, and in both cases a captive population
apparently couldn't remain viable without this artificial filtration.

What does all of this have to do with the evolution of language? It
seems likely that conversation and argument, in the broadest possible
sense of the words, are one place where culture is managed in this way.
Much of our weeding and planting, I've been arguing here, takes place
while making things explicit in conversation (as Dr. Tulp is doing in
the picture used as the frontispiece for this book). Since conversation
is a use of language, this, if true, would have something to tell us about
its adaptive function. It would suggest that one of the things for which

language evolved, one of the things for which it's an adaptation, is the management of culture. If culture can be thought of as domesticated, in Darwin's sense of the word, if it is like a flock of sheep, then words, as I argued in chapter 1, are the sheepdogs we use to manage it.

This is an interesting idea, but it needs a little more development. Instead of starting my discussion by looking at conversation in general, I'd prefer to begin by investigating a particular, specialized type of conversation, one that seems especially important as a locus for cultural selection. The kind of conversation we call "teaching" may have played a special role in our evolution and in the evolution of our ability to maintain, manage, and use culture. To explain what that role may have been, I must first introduce György Gergely and Gergely Csibra's theory about the existence of a human "didactic adaptation." Then I'll consider the broader evolutionary implications of this theory by examining Kim Sterelny's "social learning hypothesis" regarding the overall course of human evolution.

7

The Cumulative Consequences of a Didactic Adaptation

LEARNING RECURSIVE TASKS

It's hard to see how a human community could let its newer members try to master, by means of simple imitation, a dangerous skill like using fire, which has the potential to injure and kill many more people than just the learner, without occasional intervention by a more experienced fire user. If modern human children can't be trusted to play with fire without adult supervision, it isn't likely that our earliest ancestors would have been more prudent. Certainly a chimpanzee, of any age, couldn't safely tend a fire.

Managing a fire has another feature that would make it difficult for a group of chimpanzees to accomplish. It requires considerable fore-thought; for example, enough dry firewood must be gathered each day to keep the fire going at night. The time when the wood is gathered isn't the time when it's needed, so there's no immediate reward for the activity. It's something that must be done so something else can be done. The reward will come later, after a sequence of other steps.

Making and using stone tools is an activity that presents similar problems. Animals may not always be willing to die at a site where good stone is available so they can be butchered more easily. The place where the stone is acquired and the place where the tool is used may be far away from each other. Before the tool can be used, a long series of actions must be undertaken to find the right stone, modify it into the right form, and transport it to the right location.

Gergely Csibra and György Gergely (2006, 2009, 2011; Gergely and Csibra 2006) argued that it's very hard to learn a skill like the manufacture and use of stone tools by means of one-sided imitation or emulation. The imitator or emulator must know which of the model's many actions he ought to emulate or imitate, and picking up particular stones at a location far away from the source of food may not seem very rewarding. If the stone tool is to be used to make a wooden tool that will then be used to get an intrinsically desirable thing, the problem is even worse. Any cultural behavior in which a tool must be created to make a tool, or a behavior must be performed to enable another behavior, has the same problem of separation between the actions required and the reward, a quality that makes it difficult to learn by means of pure, chimpanzee-like emulation or even simple imitation. Only a learning system in which the learner at least occasionally learns in a way that involves the intervention of an experienced practitioner will suffice.

The patterns of wear on the edges of Oldowan tools, the oldest stone tool technology we know of, do seem to indicate that some of them were used for woodworking (Keeley and Toth 1981). So apparently one of the things for which early stone tools were used was making another kind of tool. But learning to do this must have been difficult because of the delayed nature of the reward and the many apparently unconnected steps needed to obtain this reward. Not everyone would have mastered all the steps. The differences in fitness that these different outcomes produced would have selected our ancestors for the ability to learn and teach recursive tasks, tasks in which an action must be taken to allow an action to be taken to allow an action to be taken to reach a goal.

A chimpanzee's efforts at emulating the success of some other chimpanzee, Gergely and Csibra (2006) point out, are invariably triggered by the immediate presence of the object of desire. There are termites to be had if only she can figure out how to get them. Given the task, defined by the presence of the tantalizing object, a tool must be found that will allow access to it. Did the other chimpanzee use a twig? Then perhaps there's some way of fiddling with a twig . . . ?

The chimpanzee doesn't keep the tool once the desired thing is acquired or the goal disappears from sight. Her ability to conceive of objects as tools is unstable, fleeting, and dependent on the situation in which she finds herself. She seeks a tool for the immediate task, not a task for the existing tool. The attribution of purposes to objects, thinking of things as being *for* something, is less important than the discovery of objects for purposes, thinking of goals as being reachable by some means.

In attributing purposes to the actions of other individuals, chimpanzees seem to assume a goal on the basis of whatever seems tantalizing to them in the immediate environment. They then try to construe the actions as a means to acquire it. Human children, in contrast, apparently assume that the action must be for something and assign a goal to it by trying to figure out what outcome, intrinsically desirable or not, it would be an efficient means of achieving. If the goal is immediately clear to them, they may come up with their own action to achieve it; they may emulate. But if they aren't sure what an action is for, they will imitate it, in order to bring about whatever the mysterious goal might be (Csibra and Gergely 2006; Gergely and Csibra 2006; Meltzoff 1988). The chimpanzee holds the goal stable and may reluctantly imitate an action if he can't figure out how else to get to it (Horner and Whiten 2005), but a human child holds the action fixed and will imitate it unless he sees what its goal must be and another obvious way of achieving that goal. Csibra and Gergely argue that the human approach to the world involves a kind of reversal of perspective, a tendency to ask what things or behaviors are for and to keep an open mind about objectives, instead of asking which thing can be used to attain an already assumed goal. We see the human artifacts we encounter as reified abstract purposes—a chair is for sitting, a knife is for cutting. This reversal of perspective is required for learning and performing complicated, recursively structured tasks. Learning the skills for making and using tools in a completely different way seems to be part of the human adaptation.

Any behavior that's in some sense conventional, Gergely and Csibra contend, is impossible to learn through the kind of emulation we see in chimpanzees. A convention of doing things in one way rather than another is necessarily somewhat arbitrary because the other way of doing them also exists. Emulation may result in learning either way of doing things. Only some form of imitation can teach us that the conventional English word for a stone is *stone* and not one of the millions of other things it might just as easily be. A move from learning by emulation to learning by imitation thus is an indispensable precondition for the emergence of Lewisian conventions, which for their very existence depend on our desire to do what others are doing, drive on the right, wear a tie, or say the word *stone* if, and only if, that's what's done in our community.

Even mere one-sided imitation by itself, however, doesn't seem adequate to solve the problem of learning recursively structured tasks like making a tool to make a tool, because of what Csibra and Gergely call the "teleological opacity" of the task and the associated problem of relevance. If we don't know why an imitable action is being performed in the first place, then we don't know when to reproduce it, what its range of application is, what sort of things should be treated in this way under which circumstances, and why. We might easily reproduce the right actions or use the right tool on the wrong occasion.

To modify or use the right kind of object or take the right kind of action in the right kind of situation, we must understand how the model that we're imitating divides the world into different kinds of things. We must somehow find a way to share the model's system of classification, even when it involves knowledge of features or aspects of things about which we as yet know rather little. Not only physical objects but also situations or occasions must be grouped into shared kinds in order for the right object to be employed on the right occasion. Certain analogies between things or situations must become conspicuous to us, and they must be the same ones the model regards as salient. We must learn to share the model's perception that situations in which the "British" are "invading" "by land" form a natural and important grouping.

Those situations that call for using tool 1 rather than tool 2, an awl rather than a scraper, a knife rather than a chair, have something in common, so the required kinds genuinely are "natural" in that genuine differences in nature are motivating them, but discovering which differentiating features separate one situation from the other isn't most efficiently performed by every individual in every generation repeating the discovery of these natural differentia from scratch (though this is the way that chimpanzees do things, presumably for the safety reasons explained in chapter 5). A culturally transmitted web of entrenched kinds of things and kinds of occasions for their use can evolve incrementally as individuals in each generation refine the received classification into a more useful one, but if each individual has to rediscover the whole scheme of classification for himself or herself, without any assistance from others, then this gradual and incremental evolutionary improvement—the sort of process that gives us every *other* complex and delicately optimized system that we find in nature—can't even get started.

This is just exactly the problem of "teaching and distinguishing natures" that Plato (1961) tells us, in *Cratylus* 388c, words are used as instruments to resolve. Later dialogues like *Sophist* and *Statesman* and Aristotle's *Categories* also seem to be attempts to explain the existence and character of these aspects of the human experience (Mann 2000), although the participants in that discussion and its sequels in the Western philosophical tradition have had trouble deciding whether the natures that needed distinguishing were actual things in the world or mere features of our languages or minds (Porphyry [270] 1975:1).

Csibra and Gergely (2006, 2009, 2011; Gergely and Csibra 2006) argued that humans have a specialized cognitive faculty for conveying this sort of information from generation to generation, for teaching generalizable knowledge, and for learning such knowledge from a teacher. Given the problems I've just described, it's hard to see how a set of human ancestors who didn't teach at all could have competed with a group that did, so it's difficult to disagree with the broad outlines of their hypothesis, even if some of its specific details still are controversial.

Perhaps the reader can remember, from childhood, the almost hypnotic delight of being shown by an adult how to do some small thing with his or her hands. Or try the experiment yourself, patiently showing a little child how to perform a simple trick involving several steps, and see if he doesn't like it as much as a dog likes fetching a stick. When human children are fortunate enough to be the recipients of this kind of sustained attention from an adult, they don't seem to care *why* they're supposed to learn how to do whatever it is, they just want to do it properly so they can please and impress the adult and receive praise for mastering the skill. They trust the adult to teach them things they need to know and often are able to suspend their quest for more immediate forms of gratification in favor of the uniquely human pleasure of receiving a piece of cultural knowledge from a benign and supportive source. (If not, human adults are often willing to act as attention police, in ways that would never occur to a chimpanzee.)

Gergely and Csibra pointed out that the only time behavior is *worth* imitating is when we don't know its exact purpose but are confident that it has one. If the point of an action is immediately obvious, we can usually come up with our own action to achieve the same end. We can emulate, like a chimpanzee. But if it's part of a mysterious adult ritual that, after many surprising twists and turns, unexpectedly culminates in the acquisition of the object of desire, we have no hope of learning how to attain that object unless we engage in blind imitation, which human children very often, and chimpanzee children very seldom, are inclined to do. Blindly imitating actions whose purpose we don't know, in the naive faith that there will be a happy ending, is a considerable investment of effort in what may be a pointless or unsafe use of time. To make it worth doing, we must presume that we're being led by a benign teacher into a useful though abstruse set of practices.

David Hume made some rather strong pronouncements in *An Enquiry Concerning Human Understanding* ([1748] 1993) about the sources of human knowledge:

> All belief of matter of fact or real existence is derived merely from some object, present to the memory or sense, and a constant conjunction between that and some other object. Or in other words; having found, in many instances, that two kinds of objects—flame and heat, snow and cold—have always been conjoined together; if flame or snow be presented anew to the senses, the mind is carried by custom to expect heat or cold, and to *believe* that such a quality does exist, and will discover itself on a nearer approach. . . . All of these operations are a species of natural instincts, which no reasoning or process of the thought and understanding is able either to produce or prevent.   (30)

In this story, as infants we encounter objects and situations already grouped into kinds, between which we notice associations, with no prior process of learning what kinds of objects and situations there are and which things fall into them. Either they're simply there in nature, or like Aristotle, Hume supposes that they're innate in every human mind, or in some undescribed process each human individual must work out exactly the same set everyone else has for himself or herself. Having somehow already classified everything in the world in more or less the same way as everyone else does without any help from anyone, we are, as little children when the story begins, sitting in a corner somewhere, watching it through a telescope, and taking note of associations as they come and go. What's missing from Hume's story is all our interactions with adults and older children, who seem to be plausible sources of conventional ways of classifying things and theories about recurring causal relationships and about which sorts of analogies between situations are useful. The sheer amount of information we acquire and the fact that we end up knowing many useful things without always being able to justify our beliefs, that we end up knowing that Kamchatka exists, that you shouldn't play with matches, and that 7 is the square root of 49, seem to argue against the Humean story in its purest and simplest form.

Humans learn by trusting others as well as their own senses, even though our trust in others is often misplaced. It's true that I can't decide whether to believe that the sun will rise tomorrow, but I *can* decide whether to believe Sidney when he tells me he's in Brooklyn. This difference exists because we get a lot of information from other people, who can't always be trusted to tell the truth. Although they tell the truth less often than our senses do, that doesn't mean that they *never* tell the truth. Even the indispensable use of fire can only be properly learned from someone we can trust (Wrangham 2009), someone who won't omit crucial warnings, someone who will watch over us while we try, on our own, to apply the knowledge to real things, and clout us on the ear in the correct Gricean spirit if we're about to burn down the whole forest. Helpfulness and trust seem to be central parts of the human adaptation.

Csibra and Gergely (2009) use an example to illustrate what we tend to learn in this way. Suppose you're in a foreign country. You see a man pick up a bottle, turn it upside down, twist its cap three times to the left and once to the right, turn it right-side up, take off the cap, and drink the contents. What should you infer? Is this good manners, a necessary precaution, or what? Supposing the man's goal to be the obvious one of taking a drink, what did all the turning and twisting contribute to that, or was another end intended? His actions, without any accompanying explanation, are "teleologically opaque."

Is it just this one bottle or just this one kind of drink that the behavior is supposed to be used with, or is it good manners or mere prudence when drinking any sort of drink in this country? How *generalizable* is the knowledge being acquired? What is the scope of application of the practice?

Another question left unanswered by the observed behavior is whether the procedure and the need for it are part of the community's store of common knowledge, of conventional wisdom, of generally accepted and acknowledged truths. Would anyone in the community expect everyone else in the community to expect them to open certain kinds of bottles in this way, or has this man invented his own distinctive method of opening bottles, which hasn't yet been accepted by the world?

If you ask him, though—even if you lack a common language, you may be able to convey your puzzlement to another human being by means of pantomime—and he demonstrates the method to you, you can find out much of this missing information. He can highlight the goal of the action: mmm, delicious liquor, makes you drunk. He can convey the reason for his unusual actions; perhaps in this country, liquor bottles have childproof caps that require this sort of complicated procedure to open. If he demonstrates it for you with the apparent intent of teaching you how to do it, you might assume that the information is generic, that you're likely to encounter other bottles of this kind, and he may even be able to tell you which kind of bottle this is. If carried out competently, the teaching or demonstration also seems to suggest (to a human, or at least a human child, who tends to assume the presence of goodwill in a reasonably familiar adult unless there's a reason to suppose it's absent) that this is "common knowledge" and not this man's specific obsessive-compulsive disorder. But if he seems obviously insane or even just incompetent in what you take to be basic cultural practices, felicitous speech acts, and so on, the method may not be worth learning after all.

Csibra and Gergely's (2009) hypothesis is that we humans have cognitive adaptations for this specific type of communication. From a very young age, we're sensitive to whether or not we are the ones to whom an attempt at communication is directed. We try to interpret communications directed to us as references to things in the immediate or (later) distant environment. When someone points at something, we try to figure out what she's pointing at. As I mentioned earlier, chimpanzees do no such thing, even though they may take an interest in what another individual is reaching for.

Dogs seem to understand informative pointing (Hare and Tomasello 2005; Tomasello 2008:42–43), perhaps because some of their most recent evolving took place in a society of helpful humans. This suggests that the changes required—a different set of assumptions about motives and different social emotions—couldn't have been difficult for humans to evolve, given the right set of ecological circumstances. If this is genuinely a new ability in dogs, if it can be demonstrated that wolves *do not* understand indicative pointing (which is controversial [Hare et al. 2009]), then it means that the ability, or a reasonably good surrogate, evolved in merely a few tens of thousands of years.

We also expect demonstrations to be performed for our benefit, and we have a tendency to interpret information given in demonstrations as generic and common knowledge about kinds of things or situations.

How important are these mere differences in disposition, in the transmission of human culture between generations? We might imagine trying to teach a chimpanzee how to use a fork. The difficult thing would be getting him to pay attention to the fork, our demonstration, what we were trying to accomplish, or what his role in the project was supposed to be. What would probably interest him more is the food on the table, whether he could have some, and why he couldn't just take some now with his hand. What would stop him? Who else is in the room, he might wonder, looking around, anyone sexually interesting or dangerous? (There could be genuine opportunity costs associated with letting us persuade him to concentrate solely on the fork.) But the fact that we expected him to attend in a sustained way to what we were attending to and to do something we had thought of for him to do—modifying the world of which we both supposedly share a picture and thereby gaining useful and generalizable information about kinds of things and kinds of occasions for their use—might never occur to him. What's conspicuously missing in a wild chimpanzee is the idea of a common interest in a joint project of managing shared attention in a shared world, or the idea that there even could be such project, or that anyone will be annoyed at him if he doesn't participate in it, or that he should care if they are.

The chimpanzee doesn't know, or at least doesn't care, that our communicative efforts are directed specifically to him; he doesn't expect us to try to get him to join us in attending specifically to the fork or, even worse, its use; and he has no expectation that he will receive generalizable or useful information in this way. For him, the fork may not even be a fork, it may just be this little metal *thing* we handed him for some unknowable and ultimately uninteresting reason. He can probably learn that there are forks in the world, but if he does, he'll learn it by figuring out for himself what he thinks they're good for.

Isn't the chimpanzee curious about why we gave it to him, about what we're trying to show him? No, not really. He isn't stupid or deliberately being uncooperative; he simply has no idea of what we're trying to do and probably cares even less. How could he know, and why should he care? *He* has no ultraspecialized didactic adaptation. He isn't subject to any social contract; he isn't obliged to pay attention. Instead, he really does learn about the world by keeping track of associations. Like the players in Ken Binmore and Larry Samuelson's (2006) evolutionary game of monitoring the environment for possible cues for coordination, he and his lineage have ended up paying only the risk-dominant amount of attention to such cues, which isn't much. He seems to have a hard time even imagining a move to the payoff-dominant equilibrium. The detailed nature of the actual mechanism by which attending closely to our instructions about using the fork would improve his personal situation isn't intuitively clear, even to humans. (Otherwise there would be no need for this book.)

Items of common knowledge, in David Lewis's ([1969] 2002) sense, widely acknowledged truths, conventional wisdom, are the things that everyone knows everyone else knows everyone knows, things that are generally admitted and therefore can be relied on when seeking coordination. Genericity, too, being an instance of a *kind* of thing, when the "kinds" involved are those acknowledged by a community, involves the Lewisian recursive replication of mental states. They're kinds at least partly by convention. We take them to be kinds because everyone else does, and we don't want to be the odd ones out. We don't want to be the only ones who don't know what a wood hoopoe is, who mistakenly identify some other bird as a wood hoopoe, or who don't know which foods are considered unclean, who gleefully eat the wrong kind of grub, who don't know how to gather firewood and bring in lots of useless green wood, who don't know how to find flint not chert, who don't even know how to tie a bowline knot. As children, we don't want to be seen as babies, who don't know *anything*.

To the extent that a kind like "flint" or "firewood" is a kind by convention—to the extent that our conformity to the regularity of treating it as a kind is a consequence of our desire to classify things in the same way our teacher does in order that we can perform the taught task in a satisfactory manner (Kripke 1982)—some crude form of conventions about classification had to be present in human societies almost from the very beginning. Indicative pointing itself is Gricean, because the pointer is mostly just telling us that he's telling us to look at that thing. If what's being pointed to is a *kind* of thing—"Get me more of these" or "No, not that, one of those!"—then the problem of figuring out what everyone knows everyone knows "firewood" is, or what everyone knows everyone knows "flint" is, or what everyone knows everyone knows a "hammer" is, also comes up right away in any collaborative effort that becomes at all complicated or in any attempt to teach anyone anything.

Does the fact that the kinds exist by convention mean that their names refer to nothing real? No, their human owners select incremental variants of these conventions to use or discard in an endlessly iterated effort to come up with more predictive ways of describing the world. Consequently, by now many of our kind-terms are as good at referring to "real things"—things worth talking about—as greyhounds are at catching rabbits or as bloodhounds are at following a scent. Presumably that's why so many of our inferences from evidence actually work. The entrenched kinds or categories involved (Goodman [1955] 1983) and the nuances of meaning associated with them are the survivors of a long process of "unconscious artificial selection," and by now they're as optimized as a sheepdog or a maize plant.

Of course, in domains where the incentive for experimental learning has long been absent and everyone really is just imitating imitations of imitations, the categories that the members of a community believe in may well have drifted very far away from any underlying reality. In principle, though, words that are still closely tied to real things, and frequent practical experiments involving them, can accurately pick out useful similarities in the world.

We have these adaptations for participating in a communication system, teaching and learning from a teacher, a parent, or an older peer, whose function seems to be transferring pieces of Lewisian common knowledge. These adaptations let us receive from and convey to others knowledge about what everyone knows everyone knows, about how to do things in ways everyone knows everyone knows are the right ones, about what everyone knows everyone knows is good and bad behavior, what everyone knows everyone knows words mean, and so on.

What the teacher creates, or maintains and preserves, is common knowledge. This means that unlike emulation, which is useless for this purpose, the system by which humans learn is perfect for transmitting knowledge of Lewisian conventions. It's perfect for teaching us what everyone knows everyone knows about how to tie a tie or drive a car on a public highway or exactly what the word *flint* means or under what circumstances one is supposed to hunt rabbits instead of deer. Teaching us the conventional wisdom of our community is what the adaptation is designed to do, so that we can participate in the coordinated activities that it's built around. (Remember how difficult that can become, without precedents like the one telling us we should drive on the right.)

For the purpose of adaptive filtration, one virtue of this system is that in a well-run society, the teacher is probably experienced, having, we can hope, already subjected his or her version of the cultural behavior to trials in the field, which may have led to culling the inferior version that he or she started out with and replacing it with something incrementally better, or at least more standard. Imitative learners—which is what we all must be when we're young—are put in direct contact with seasoned experimental learners, thereby saving them from the risk of imitating another imitator or persisting in a self-generated error.

Because of this practical experience, the teacher may also be able to spot corrupted or maladaptive versions of associated behaviors or beliefs, which may cause problems in learning the new skill, convention, or behavior, and to offer a better, more conventional, or more serviceable replacement for those, too. At the very least, the teacher has the advantage of not being the student and therefore being less likely than the student to be in error in the exact same *way* as the student is. The teacher may be wrong about many things, but if the student chooses well, these will not be the same things about which the student is wrong. Teachers who teach useless versions of skills may fall out of favor, so there's adaptive filtration by the pupils as well, which might help filter out skills that no longer are suitable for the current environment, as well as ones that have become corrupted in transmission.

A crude or an imperfect capacity for a recursive theory of mind might occasionally have appeared as a sort of side effect of selection for the ability to engage in the recursive manufacture of tools. Making a tool to make a tool to make a tool and thinking about what someone thinks about what someone thinks are actions of the mind that have a certain similarity. But as Gergely and Csibra point out, there's another, more convincing connection between these two capacities, a functional link. Both these things—how to make a tool to make a tool, and what she knows that the student doesn't know and doesn't know that she knows— are things a *teacher* must think about at the same time and in the same situations. The teacher must think about what the student doesn't know that the teacher does know about how to do or make the thing needed to do the thing needed to do the thing the student does know that he or she wants to do. The teacher also must persuade the student to pay attention to the right things, the same things the teacher is paying attention to. So it seems to me that it isn't right to think of two separate and unconnected adaptations, one for recursive toolmaking and one for a recursive theory of mind, which happened to be evolving at the same time, in the same small groups of individuals, in an unconnected way. Functionally, from the very start, the two things must have been fused together into a single twisted mass of knowledge about knowledge about knowledge of steps needed to take steps needed to take steps, a single complex didactic adaptation, which also acts as an adaptive filter for the items of culture it transmits.

THE EVOLVED APPRENTICE

If we included the idea of a didactic adaptation in our story, what sort of overall macrotheory of human evolution would we have? Kim Sterelny (2012) looked into this question and came to very interesting conclusions.

Sterelny was unhappy with what he calls the "social intelligence hypothesis" in our theory of the evolution of modern humans, and wanted to replace it with a "social learning" model. He objects to various versions of the hypothesis (Dunbar 2003; Humphrey 1976; Miller 1997), but he points out that the feature they all share is Rousseau's assumption that the division of some preexisting pool of resources and tasks was the main problem facing our primitive ancestors. It's assumed that their fitness depended mainly on not being cheated or outmaneuvered, on obtaining their fair share. The problem of learning how to obtain the resources in the first place apparently was of secondary importance.

Sterelny points out that evolutionary game theory encourages this "Machiavellian" perspective on human evolution, since it's mostly about strategizing to acquire a greater share of an abstract good, which is assumed to come from more or less nowhere, in competition with others. (Sometimes, as with Rousseau, there's a thin story about hunting deer, but the focus is never on the inherent difficulties of this project; it's always on strategic calculations about how to get the biggest payoff for the least effort when others are trying to do the same.)

In this "social intelligence" approach to our evolution, the main problems facing early humans are thought of as having been essentially political rather than practical. To deal with these basically unchanging fairness issues, we're supposed to have evolved a set of fixed-purpose "social intelligence" modules whose function is to facilitate Machiavellian scheming, to police obligatory altruism, or to do both at once. That—the ability to scheme against our peers—rather than our ability to have a culture, is taken to be the central human adaptation.

This seems like a tough-minded, realistic approach to human nature, but I agree with Sterelny that it doesn't adequately explain human uniqueness. Other primates have complex, Machiavellian social lives as well. Humans, in particular, seem uniquely inseparable from the culturally transmitted extended phenotype (Dawkins 1982) that we call technology. A modern human with no technology at all and no knowledge of technology, no concept of using fire or shelter or butchering tools, will die very quickly in almost any natural environment. We simply can't survive on raw food for very long (Wrangham 2009). To live, we humans must cook, and to cook, we must have fire, and to have fire burning where we sleep, where any member of the community can easily get at it, we must teach, in the broadest sense of the word *teach*, which again may be simply giving someone a Gricean thwack on the head at appropriate moments in the development of his competence.

Is the human affinity for producing and conserving powerful technologies really just a by-product of an adaptation for scheming? No doubt social complexity was one of the influences on our evolution, but to think of it as the only important one makes too much of human nature an accident. Given the strong emphasis on egalitarianism in most hunter-gatherer societies (Boehm 2001), life in a historically typical human society might actually be much *less* demanding with respect to these kinds of skills than life in many other, more hierarchical and fractious primate societies. Dogs are a sort of mirror. They show us what our recent ancestors have tended to reward—focusing on our faces, making direct eye contact, attending to our instructions, expecting to be helped, and desiring to be helpful—which can tell us what those ancestors were like. They thrive in human societies, from a Darwinian point of view, despite their lack of Machiavellian modules, by following a simple strategy of unquestioning loyalty to their human families and barking at strangers. Shouldn't the existence of some sort of crude, bonobo-like social contract have made social interactions easier, not harder, to manage?

Sterelny thinks we can find a better explanation for the unique features of human minds. His approach emphasizes the cultural transmission of productive skills.

Until a few million years ago, our ancestors probably lived like chimpanzees, scrounging for whatever fruits and nuts and small amounts of meat they could acquire by themselves, without much technology, and consuming them on the spot. But at some point, they turned into what he calls "cooperative foragers," who used a larger variety of harder-to-process and harder-to-acquire foods in ways that required complex, changing skills and cooperative effort. At some point, they started hunting big game, using fire to cook food, and digging up and processing roots.

The crucial requirement of this new way of life was the transmission of skills from generation to generation. Children and adolescents had to become more interested in what their elders were doing, and their elders had to become more tolerant of that interest. We observe this sort of interest and this sort of tolerance in New Caledonian crows, who have a fairly complex set of tools for fishing insects out of holes. (For an example of the large literature on these behaviors, see Holzhaider, Hunt, and Gray 2010.) It seems obvious that making stone tools, using fire, and hunting big game require the kinds of skills that can be acquired only in this way, but Sterelny points out that the collection and processing of wild plant foods other than fruits, leaves, and bark offer similar opportunities for developing culturally transmissible cooperative skills.

He sees this primitive imitative relationship as evolving into a form of apprenticeship, with younger individuals performing relatively simple, undemanding tasks in return for the privilege of being shown how to carry out more complex tasks. This interaction need not involve altruism on anyone's part—the master benefits by having lots of simple tasks carried out by someone else, and the apprentice benefits by acquiring a craft.

Certainly, people who live in a traditional way do seem to have an impressive level of skill and knowledge regarding the environment they live in, one that seems impossible for any individual to achieve entirely on his own, like a chimpanzee, displayed in activities like the hunting of game and the gathering and processing of plants. Tracking, alone, is an incredibly complex skill. If you're really going to learn how to do it well, someone must point out the signs somehow and perhaps give you some minimal kind of explanation while you listen or watch attentively. Then you must try to apply what you've learned, often earning the silent or highly vocal derision of the more skilled people around you when you mistakenly identify the footprint of a fox as belonging to a wolf or suggest that a track is from last night when it's obviously from the previous day.

It's deeply embarrassing to be shown ignorant of something that's supposed to be common knowledge, because it means that you aren't party to the conventions of your community, in the sense of being able to conform to them. If you're too foolish to know the difference between a wolf's foot and that of a fox, how can you be considered an adult; aren't you still a little *child*? To misapply a crucial name is to demonstrate that you're not yet a full member of the community of speakers of your group's language.

The archaeological record contains many sites where adults made flint tools, and next to them, children used the same stone to make inferior copies (Grimm 2000). Were the children embarrassed by their mistakes? Better for them if they were. Did they crave the adult's approval? Better for them if they did. But as far as we can tell, a chimpanzee never feels or does anything even remotely similar to this.

Also frequently found in the archeological record are tools that were crudely started by an unskilled individual and then deftly finished by a more skilled one. Sterelny connects these relics to the way in which the production of stone adzes is organized in modern New Guinea (Stout 2002). Would-be adze makers must serve long apprenticeships, during which they start out performing simple tasks like roughing out blanks and polishing completed adzes and, in return for their labor, are gradually initiated into the complexities of more advanced craftsmanship. Many hunter-gatherer groups integrate children into adult work activities, giving them child-sized versions of adult tools and informal but explicit instruction in what to do with them (Haagen 1994; Hewlett et al. 2011).

A human apprentice doesn't simply *imitate* the master craftsman or accomplished hunter to whom he has attached himself. The master doesn't simply go on about his business, confident that the apprentice is learning the trade correctly just by watching him. Sometimes the master frowns or makes a sarcastic comment. Sometimes the older apprentices smirk, and the young apprentice is quite likely to care if it's he at whom they're frowning or smirking. Human children care what adults think of them in ways that find no obvious echo in chimpanzees, outside the bond between a mother and her child that we have in common with all mammals. To care what people think of you is to see yourself through their eyes, and seeing yourself through the eyes of others seems like a possible evolutionary path to common knowledge and Lewisian conventions, to everyone knowing what everyone expects everyone else to expect. It's also a strong incentive to abandon or replace any learned behavior that will make the adult frown in disapproval or cause the older apprentices to smirk in disdain, and a strong incentive to attend to the things that the adult expects you to attend to. It gives you a reason not to experiment with uncooperative behavior and a reason to try to avoid making careless mistakes.

Caring what others think of us plays an important role in the transmission of culture between human generations, helping us orient ourselves to the mental efforts needed to escape from our condition of ignorance or inexperience or from having learned the wrong thing. A human cares about whether he or she does things well or badly, is proud to know how to do something well, and is ashamed to do it in the wrong way. I'm not sure that chimpanzees feel anything like that. They may be happy to finally know how to get nuts, but I think there's good reason to doubt that they're *proud* of the hard-won skill of cracking nuts with a hammer. I don't think they'd ever look around to see if anyone had noticed how well they were doing it. Instead their focus would be on the nut.

All this suggests that the human role of teacher isn't a recent or a trivial invention, but is one with deep roots in human history and human nature, a fairly central part of our distinctive adaptation. To allow recursive tool use and cumulative culture, we would have to have had a crude form of teaching, "teaching and distinguishing natures" and discouraging errors, almost from the beginning, almost as soon as we started to differ from chimpanzees.

The role of domesticator of culture is one that the teacher, or the apprentice's master, is well placed to carry out. Teachers can weed out errors in cultural behaviors or in our way of understanding our conventions about meanings by explicitly discouraging them. But they can also influence what's regarded as conventional more subtly, because of their superior ability, as sources of authoritative common knowledge, to make their own private judgments into public focal points and binding conventions. ("This is the way we make adzes here.")

Thomas Schelling (1966) argued that a mediator can often decide the outcome of a negotiation by making one of the possible outcomes the most obvious one, and thereby making it impossible for the parties to the dispute to agree on any other outcome. The existence of the sorts of social emotions I've been describing must make it possible for teachers and other people in vaguely similar roles—chiefs and judges and priests and poets and parents—to sometimes play a rather similar coordinating role with respect to conventions, filtering out the ones they don't like and promoting the versions they prefer. Since we all choose words in conversation, we all have a role in their selection, but the teacher, as Socrates argued, must play a special role in this process.

Parents and other family members play a similarly crucial role in the process of language acquisition earlier in life, passing on binding conventions about meaning simply by making them the obvious ones and by making learning them possible at all. Unfortunately, parental behavior is one of those things that just don't fossilize well. It's much easier to know how tools were made in the distant past than it is to know how children were raised by the makers of those tools. If we want to know how human parents, other adults, and older children interact with human children and whether a child's acquisition and retention of language is assisted by human choices, if we want to know whether the child or the parent or both make a concerted effort to weed out inadequate efforts to participate in linguistic conventions and encourage better ones, we have no choice but to look at the present, at modern human parents interacting with modern human children. We must bear in mind that what we're seeing is one end, the most recent stage, of what probably has been a very long and elaborate evolutionary process, with many steps and stages and false starts along the way. Still, these are the only facts available to us, so the present is where we must go to retrieve them.

Now that I've raised the subject of the way in which language is acquired by modern humans, as much as I'd like to linger there, it's time for me to leave behind the Paleolithic and cultural evolution in general. If the conclusions I've drawn from our extended tour of the distant past are of any use, they should be reinforced, and not undermined, by a look at the language-learning practices of modern people. If Lewisian linguistic conventions are an example of domesticated culture, that fact should show up in the way they're implanted and cultivated in the minds of new human individuals. So in the next chapter, I'll move back to the modern world and examine the early part of language acquisition as it happens today.

But before I do, there's a question that won't wait any longer, that I must deal with at once, now that modern human language is once more on the dissecting table. Meanings, it would seem, are conventional, in David Lewis's sense of the word, and Lewisian conventions, as I've been arguing, are domesticated items of culture. But what do I mean in the first place when I use the word *meaning*? What is a "meaning" in a story like the one I'm telling, a story in which meanings are settled over and over again in conversations like the one between Dr. Tulp and his audience? How do we learn about them, and what's accomplished in these conversations? Does anything actually end up being true by definition? Once I've dealt with these subjects, I'll be in a position to think about how we learn to interpret the things that are said to us in conversation. But first I need to say something about the meaning of the word *meaning*.

8

Meaning, Interpretation, and Language Acquisition

THE "BATTLE OF NAMES"

It's time to make the transition from cultural evolution in general—which I had to consider in order to make sure that the idea of domestication was a useful way of thinking about human culture of any kind—back to my original and primary focus: the evolution of human language in particular. I began this book by asking how the words we use got their meanings. But what is a "meaning"? Over the past forty years, no philosophical topic has been discussed more thoroughly, and few have been more controversial.

In this section, my goal isn't to engage in controversy with other philosophers; it's to explain, for readers who may not be professional philosophers, what a philosopher might mean by the word I'm using and then to make one fairly simple point about what that implies about the role of argument and conversation in the management of human language and human culture. My objective isn't to redefine the word *meaning*. Philosophers don't universally agree with one another about what it means, so I can't give an uncontroversial definition, but the account of human language I'm giving here is built around one of the more conventional philosophical interpretations of the word.

In explaining what I mean by it, I'll try to stay within the boundaries of the account given in Hilary Putnam's "The Meaning of 'Meaning'" (1975b). What he says there seems right to me, and his account of meaning is common knowledge among philosophers, so reliance on it makes my own story about the evolution of meanings easier to explain. Readers who disagree with Putnam's account can refer to his many defenses of his own position, so there's no need for me to fight those battles again here. Just as with all the many literatures I've touched on in this book, I'm in no position to review the literature on this subject, because it's huge and complex and because other people have already done a better job of that than I could. Anyone who wants a more comprehensive discussion of Putnam's views might begin by looking at Andrew Pessin and Sanford Goldberg's *Twin Earth Chronicles* (1996).

I can't recapitulate Putnam's whole argument here, but for those not familiar with it, I'll outline some of the high points. His basic complaints about more traditional theories of meaning are that they portray meaning as being all in the head of one individual. They treat it as essentially a psychological, not a social, reality, and they ignore the role of things themselves in giving our statements their meanings: "The grotesquely mistaken views of language which are and always have been current reflect two specific and very central philosophical tendencies: the tendency to treat cognition as a purely *individual* matter and the tendency to ignore the *world*, insofar as it consists of more than the individual's 'observations'" (Putnam 1975b:271).

To show that meanings aren't just in our heads, Putnam invites us to imagine that we discover another planet, Twin Earth, which is exactly the same as ours in every way, including the widespread use of what seems to be English, except that everything on Twin Earth that seems to be water and that the inhabitants refer to as "water" is in fact another superficially indistinguishable fluid made of a completely different kind of molecule: XYZ instead of H_2O.

Suppose both John and Twin John go to the corresponding beaches on the two planets, dip their toes in the thing they call water, and exclaim, "Gosh, the water's cold today!" Do they *mean* the same thing by the word *water*? No. John is referring to the H_2O in the ocean on Earth, and Twin John is referring to the XYZ in the ocean on Twin Earth. To say "On Twin Earth, XYZ is water" or "For Twin John, XYZ is water" would be to say something inaccurate, possibly leading to the incorrect expectation that during a water shortage, we could go to Twin Earth and scoop up some XYZ. In that case, we wouldn't have obtained what *we* call water. John and Twin John are in the exact same psychological state when they say the two sentences, since they don't know anything about the difference between water and XYZ, the two liquids look the same to them, and their sentences are identical. But because the world around them is different, they mean different things. The extension of the term, the set of things in the world to which the speaker is referring, is different in the two cases even if the two speakers' state of mind is exactly the same. Presumably, if they did enough work or consulted the right experts, they could find this out, could even find out that one of them was actually referring to XYZ and the other to H_2O. We would be wrong, therefore, to believe that they are saying the same thing. In the end, when everything finally was clear, they themselves would conclude that they hadn't been saying the same thing, that on Earth *water* is the word for H_2O, and in the slightly different language of Twin Earth, it's the word for XYZ.

Many of us might not know what chemical tests to perform to decide whether something really is water and not XYZ, but we trust the experts to know the tests and to have performed them on the ocean if, in their expert opinion, that was called for, and we rely on that presumption in referring to the "water" in the ocean. We rely on what Putnam calls a division of linguistic labor. Even though John and Twin John themselves may not know exactly which other pools of liquid are pools of the same liquid as the substance they're characterizing as cold, their words presume that it *is* something in particular, just as a scientist who says "The unknown substance on Planet X is very cold" presumes that there is something that the unknown substance *is*. They're right, and in their two cases it really is something different, so they are referring to two different things, despite being in the exact same psychological state.

We're supposed to draw at least two morals from this story. The first is that many meanings are indexical, like the meanings of the words *you* and *I* and *this* and *that* and *here* and *home*. When we say the word *water* or *gold*, what we mean is "this stuff here that we all call *water* and every-thing like it" or "this stuff here that we all call *gold* and everything like it."

Terms like *water* and *gold* also designate rigidly: in any possible world, these terms always are supposed to designate the same thing, this stuff here. I can't coherently speak about a world in which gold is actually aluminum or conventions are actually kettles. But in a sentence like "The winners of the World Cup are usually the best players around," we mean instead, "whoever happens to fit the description 'winners of the World Cup' at any given moment," not the current champions. Terms like *water*, *gold*, *Richard Nixon*, and *convention* are supposed to designate rigidly, to pick out the same thing, person, or place in every possible world, whereas more flexible terms like *surgeon general* and *winners of the World Cup* can sometimes be used in ways that require choosing whoever or whatever fits the associated description in a particular world or at a particular time. Putnam pointed out that it's a mistake to forget about the indexical, and therefore rigid, character of many of our terms and to suppose that their meaning is merely the descriptions, associations, or senses that we also attach to them. When someone names something, or some kind of thing, the name continues to attach to that thing, or that kind of thing, and doesn't at some point start to stand for anything that happens to fit a supposedly accurate description of that thing. As Saul Kripke argues in *Naming and Necessity* (1980), the fact that we may believe that gold is a yellow metal doesn't mean that if we found some gold that wasn't yellow for some reason, it wouldn't be gold.

The second moral is that a society's use of language involves a division of labor and that names have histories, with some people relying on the more sophisticated knowledge of others to guarantee that things really are what they're calling them. Once someone has named gold *gold*, the meaning of the term is given by its provenance, even if the descriptions that we, personally, happen to associate with the term are faulty, even if we, personally, happen to think fool's gold is a form of gold or have managed to convince ourselves that gilded lead is gold. Not all of us are goldsmiths, but we can refer to the earrings we bought from a reputable goldsmith as "gold earrings," secure in the knowledge that he's probably done whatever is necessary to make sure they are made of that thing we all call gold. We reasonably suppose the characterization is accurate because of what we know about its provenance. Language, Putnam tells us, is more like a steamship than a hammer or a pencil; different people in a society play different roles in its operation.

When our concern is dynamics, the way that a community's conventions about meanings come and go and change and evolve over time, I believe that there's a third moral in the story as well, one that Putnam seems to assume in "The Meaning of 'Meaning'" (1975b) but doesn't explicitly discuss. The way I told the story, John and Twin John could have met, discussed which thing should be called water, and consulted experts. Then they would have had to make up a new word for one of the substances, perhaps XYZ. They could have renamed it XYZ, as we've done. Or they might have decided to call it something else, but from then on there would be two words, in the languages of both planets, one for water and one for XYZ (as there now are in our language). Both languages would have been enriched, and the enrichment would have been the result of a conversation, in some extended sense of the word *conversation*.

What were the two parties originally confused about? They were using the same name for two different things, and they didn't know it. They didn't know how to distinguish between water and XYZ, even though they are completely different. But later they discovered the difference, and then they weren't confused any longer.

If someone showed John a photograph of an ocean on Twin Earth, he once might have said, "Look at all that water!" He would have been mistaken, however, and would have been misapplying the name because the picture would be of a different kind of stuff. At that point, he was prone to making mistakes about which thing the word in his language should be taken as referring to, because he hadn't yet learned to distinguish between water and XYZ. Later he found out and realized that he'd been applying the name incorrectly.

People often misapply names as a result of incomplete information or imperfect reasoning or an imperfect language. They carelessly or sneakily call one thing by the conventional name of another. They routinely resolve these confusions in the same way John and Twin John could have, in conversation or by consulting an authority. When the evidence changes, as it would have for John and Twin John when the first astronauts from Earth visited Twin Earth, or if the existing language simply isn't yet making the right discriminations, or its terms don't yet carry the right associations, then that language must change as well, and these are the sorts of methods by which it incrementally changes.

Because this sort of question constantly is settled in conversation or consultation, we must describe a lot of things and interpret a lot of descriptions. The chemist must describe water well enough to allow us to interpret the term correctly as referring to the same kind of stuff in our own rather different environment. We must interpret his description carefully enough to make sure that what we're calling water is only what he would call water.

In interpreting names we've received from others and describing the things we think the names apply to, we may end up applying them to slightly different sets of objects than our predecessors did, or describing the same things in a slightly different way. Usually these reinterpretations and redescriptions are mistakes that cancel out each other or are mere noise, but sometimes they must be slight improvements. They even may be fairly dramatic improvements, like those associated with the discovery that water itself is actually H_2O, a discovery that would have changed many of our descriptions of it. At other times, though it's a matter of describing situations in which we once would have said that it's raining as those when it's "misting," or slightly refining our received notion of the difference between polenta and grits, or starting to interpret the term *condescending* to cover a different set of behaviors and attitudes (as Jane Austen already was beginning to do, since her use of it actually damns with faint praise the person described), or beginning to use the word *approve* to mean *like*. Whatever it is, the new way of using, interpreting, and explaining the word either will or will not catch on, depending on whether the other people in the population find it useful or attractive and on how well placed we are to make the new version seem like an obvious choice.

It is difficult to doubt that these two reciprocal processes, interpretation and description, going on around us all the time, have some effect on the community's evolving sense of what the name stands for and how those things should be described, and on the extension and senses of the name itself as it's understood in our conversing community. For names like *gold* and *water*, this process makes little difference for extensions, because the physical object itself, the element or chemical species, provides a secure, rigid, unchanging, and easily observable anchor. But when we speak of conventions, meanings, causes, species, organisms, genes, witches, injustice, truth, or even Madagascar—which, Gareth Evans (1973) tells us, started out as a name for a part of the African mainland—there often has been considerable room for subsequent improvement or amendment in what we originally meant by the word.

Where does all this leave Quine's ([1936] 2004a, [1951] 2004b) original concern, "analyticity," things being true by definition? Because of the social character of meanings, Putnam in effect replaced the term *analytic* with *obligatory*. When something is true based on the conventional definitions of words in your speech community, when something is gold as a competent goldsmith would define it, you aren't playing by the rules of that community if you treat it as untrue, if you claim that it isn't really gold, or if you present a different metal as also being gold, without at the same time arguing for a change in the convention. (Lewis [(1969) 2002:97–100] made this point when he observed that many conventions also are norms and must be understood to have normative force, even if most people in the community adhere to them for self-interested reasons.) Some ways of using the word *gold* will be greeted with strong disapproval—for example, if people catch you using it to refer to bars of gold-plated lead that you're trying to sell to them. In this sense, to say that the claim that gold is the element with atomic number 79 is analytic is just to say that using the word to refer to the element with atomic number 82 is a flagrant violation of a binding convention.

Thus the correct accusation of the person who denies what's "true by definition" isn't that he's a madman; it's that according to his peers' standards, he's either behaving badly or is confused, because he's not giving words the connotations or extensions that are normal and obligatory in his particular speech community—among botanists or mathematicians or goldsmiths or logicians or constitutional law scholars or ordinary people—and he isn't even warning them that he's doing this. In some sense, he's undermining the social contract by making language less useful and coordination more difficult. Yet the conventions of language, like any others, can become obsolete or distorted, so if he has a point, if he's doing it for a good reason, he may be doing the exact opposite, he may be making the language stronger.

What, then, should we expect to happen in the sort of world that David Lewis describes in *Convention* ([1969] 2002), a world in which everyone speaks a slightly different version of a shared language? Since people attach slightly different meanings to words, they'll regard slightly different sets of sentences as true by definition. When two members of the population meet, they may discover that they interpret different sentences as true by definition. That means that each of them may think that the other is being unreasonable or at least is mistaken, unless they're persuaded by the other's way of looking at things.

Each may feel that the other is either deliberately or inadvertently violating legitimate binding conventions regarding what words are supposed to mean. There's nothing to stop them from saying, "Of course that doesn't follow, you're giving the word the wrong meaning, you're taking it to refer to the wrong things." (Not all our bargaining about the conventions of language is tacit.) Each may believe that she's using her words correctly and that the other is using words perversely.

If despite this, they still think it's worth talking to each other, they may argue or converse until they've identified the source of the problem and harmonized their idiolects: "I see, so by *fitness* you really just mean the number of offspring that survive to sexual maturity. Well, that's a bit different, I was thinking of the word as meaning health and strength." In so doing, they will settle a disagreement about the meanings of some word or words that before the conversation, they may not have known they had. If the way they settle their disagreement is persuasive, if they come up with a clearer or more useful account of what the term *Darwinian fitness* means than people had done before their argument, the new or clarified convention regarding the meaning of the term may spread like a virus through the idiolects of the members of their community. The community's members may think that this really must have been what the word meant all along if only they'd known, or they may self-consciously prefer the new, clearer sense of the term *Darwinian fitness*.

What counts as "analytic" in one or both of their idiolects, and therefore what counts as being unreasonable, may then have changed. In far-off worlds where each of us once saw something different, we all now will see the same things, so in effect, our shared system of possible worlds has been expanded and amended, and new worlds are now available for coherent discussion. This is the evolution of language in action, a selection event in which one interpretation has outcompeted another. Many of these changes may be quite small and subtle and may be the results of tacit processes rather than explicit ones, but a long cumulative series of such subtle changes must have produced the complex and finely optimized tools for "distinguishing natures" that we all use to converse today. Because the process is cumulative, these tools are optimal in a way that may not ever be fully understood by any single user of the language.

Metaphysically necessary truths, if there are any, must have their source in the very nature of things. But analytic truth, "truth by definition" in the language of some human population, is different. It's our best collective, cumulative estimate of what is metaphysically, or at least physically or practically, necessary. It's the outcome of a collaborative, iterated process that's wiser than any single one of us, but we still have often been wrong about this sort of thing, so we need a mechanism for bringing the two sets of "truths" closer and closer together. "Analytic truths" that don't conflict with what's necessary or possible in the real world don't just fall from the sky. Work must have been required to create them—not fictitious work done by Adam at the dawn of time, but real work done by a long series of real people.

As in all evolutionary processes, it seems to me that here the behavior and its consequences are the design process for the behavior. We work out the useful truths-by-definition by repeatedly using words in certain ways and getting away with it, or by repeatedly arguing about what *is* analytic, about what words *should* mean, and about what *ought* to be true by definition, by carefully redefining the word *parallel*, all the time assuming that the right answer to the question of what it means already exists. Quine's ([1951] 2004b:35) lexicographer, who supposes he's an empirical scientist, seems to be in the grip of a kind of Platonism that afflicts many of us. He assumes that the word he's defining already has a set meaning, which he must simply discover and report; but in trying to discover it, he is actually helping decide what it will be.

Socrates referred to this general process as the "battle of names" (*Cratylus* 438d). But without our help, how are the names supposed to battle? When we disagree with one another about what words mean, about their correct interpretation, or about how to describe the things they stand for, we're taking sides in these fights, and it's we, in conversation, who ultimately settle them. Sometimes we do this explicitly, and sometimes we do it tacitly, by setting an example.

LEARNING TO INTERPRET

Now I'll return to the question that I raised at the end of chapter 7: Is a child's acquisition of the semantic part of language assisted by human choices? Is there any weeding out or pruning of inadequate performances by the child, the parent, or both acting in concert?

My own first words were *kitty cat*! I probably wouldn't have known at that point that the word *cat* also refers to lions but never to dogs. How did I go from that first step to my current practice of using the word *cat* to refer to all felines, and only felines, and the rich and nuanced set of associations I now attach to the word? How do modern human children learn the words of their first language? How do we, as individuals, first acquire the ability to participate in our culture's domesticated linguistic conventions? What role do more competent speakers play in nursing and cultivating our new idiolects as they begin to grow and develop?

In trying to answer these questions, I'll explain the rough outlines of what Michael Tomasello calls the "social-pragmatic" theory of language acquisition. Since I'm not an expert on this subject, I'll rely on Jerome Bruner's classic account of the process in *Child's Talk* (1985). Unfortunately, I'll have to skip over many intrinsically important considerations, but I'll try to explain the general idea.

Children typically begin to learn words around the end of their first year and the beginning of the second. According to Tomasello (2008), referencing Bruner, "Virtually all of children's earliest language is acquired inside routine collaborative interactions with mature speakers of a language—in Western culture, such things as eating in the high chair, going for a ride in the car, changing diapers, feeding ducks at the pond, building a block tower, taking a bath, putting away the toys, feeding the dog, going grocery shopping, and on and on" (157). Notice the very tight connection, right from the start, between collaboration, of a uniquely human kind, and learning, of a uniquely human kind.

These interactive "formats," as Bruner called them, seem to be exactly what Ludwig Wittgenstein (1953) said he was talking about when he introduced the notion of a "language game," in *Philosophical Investigations*, section 7, as "those games by means of which children learn their native language" (5). These are small, structured, real-world interactions involving coordination or collaboration, with the coordination coming at least partly from the use of language by one or both parties. These interactions are crucial to language acquisition because they're occasions for the adult and the infant to share intentions, to be part of a "we" engaged in a common activity. This, again, is something that humans do that chimpanzees do not do, apparently can't do, and have no instinctive childhood process of learning to do, so the difference seems important.

To participate in these interactions in a way that will elicit reinforcement from adults, children must learn to join them in attending to the same aspects of objects or actions or situations. This process of learning to direct their attention to those aspects of things that are being attended to by the adults interacting with them seems to be how children learn to see some "analogies" as "conspicuously salient" or as "natural" (to borrow Lewis's [(1969) 2002] description of the required result) while other candidates (for example, my own early practice of referring to both cats and dogs as *kitty cat*) are weeded out. The process of conveying to an infant the society's ways of classifying and valuing things in the world requires the active participation of adults or older children.

Even though children may learn grammar from adults who are ignoring them, both Bruner and Tomasello tell us that it's rare, at least in the beginning, for them to learn a *word* from an adult who isn't attending to the same things they are. Almost all early words are first produced in interactions involving joint attention, more often as an answer to a question from an adult or another child than as an unsolicited imitation of what he or she has just said.

It makes sense that these two things should be learned in different ways, because the syntax of a human language, though complex, is apparently more or less arbitrary, and what must be learned about it is almost completely independent of any particular physical, social, or technological context. No facts about the world have to be learned as part of the process of learning to speak grammatically; it's just a matter of toggling the switches on the Chomskian hardware into the same combination that everyone around us is using.

In contrast, learning the meanings of words like *cat, firewood, convention,* and *electricity* requires learning massive amounts of information about the contingent situation in the social and physical world that we inhabit. We must learn to interpret sentences that contain these words as referring to particular aspects of particular things or situations or kinds of things or kinds of situations, such as the unhappiness of a specific cat whose tail has been carelessly stepped on, the dryness of a specific piece of firewood, or the binding character of many conventions.

As Putnam (1975b) and Kripke (1980) argued, meanings are deeply embedded in our physical and social surroundings, so it's hardly credible that these sorts of interpretations could also be put together in our head automatically and unconsciously, without much experience or much help from more accomplished speakers. Learning syntax probably is just a matter of flipping on the right preset switches on some innate Chomskian "language acquisition device." But learning to interpret means learning to navigate in a whole world and, in a very nuanced way, learning how all the things *in* that world are classified and evaluated by the people around us, learning what's what, and what's important.

These early instances of joint attention management seem to be biased toward interactions with physically present objects, with the sphere of possible subjects of conversation gradually expanding outward as language develops and the children get better at interpretation. The game of Book Reading (which I'll explain in a moment) shows us some of the ways this expansion beyond the immediate and the physical takes place.

Bruner investigated a number of "language games" like these by watching (and videotaping) two small children interacting with their mothers. One was peekaboo, which begins to be played during the child's first year, before he (both of Bruner's subjects were little boys) has much language at all. The other game I'll discuss is Book Reading, the cooperative inspection of a picture book, presented by Bruner as a sort of extension in the child's second year of an earlier language game asking the questions "Where's the *x*?" (Where's the kitty cat?) and "What's this?"

Readers may wonder whether I suppose that adults have been playing peekaboo, and Book Reading, in particular, with children since Paleolithic times. Of course, at least in the case of peekaboo, cross-cultural variation *has* been studied. The Xhosa equivalent of "Peekaboo, I see you!" is "N-a-a-a-an *ku!*" From Anne Fernald and Daniela O'Neill's (1993) discussion of this topic, we learn that there are many similarities across cultures in the way the game is played, and also some differences. I don't think parents read picture books with their toddlers during the Paleolithic. I do, however, think it's likely, as Bruner suggests, that some game or routinized interaction that ontogenetically recapitulates the original giving of names in every single human generation is probably a common feature of human cultures. I think that every human culture must have a way for its children to learn how to interpret what's said to them.

These differences in the language acquisition games played by people in different cultures aren't a problem with Bruner's story about the role of games like peekaboo and Book Reading in language acquisition. One of his main points is that children acquire language in the course of, and partly as an aid to, acquiring a particular culture, by playing the language-acquisition games specific to that culture.

Adults seldom correct a child's grammatical mistakes, and their direct corrections of mistakes involving meanings are inconsistent and probably inadequate. But from the beginning, all reasonably committed human parents do seem to be concerned with critiquing what John L. Austin (1962) called "speech acts," the employment of words strung together in a grammatical or ungrammatical way to accomplish things in the world. From well before it's plausible that the child could have any real idea about such things, felicitous requests receive better responses than infelicitous requests. Parents want their children to become civilized human beings, and much of their inadvertent language teaching takes place while helping them do that.

What constitutes felicity and infelicity must vary greatly among human cultures, but in England in the 1970s, requests for things that the child didn't want or could easily get for herself or that treated the adult as an adversary, not an ally, or that didn't adequately specify what was being requested were less likely to produce an immediate positive response than their opposites were. One thing that we must learn in order to make our remarks worth interpreting, or to interpret remarks that are made to us correctly, is how to act in good faith, how to show a human kind of goodwill to our conversational partners.

In these interactions, children learn how to do things with words in the culture into which they're being inducted. Most of the focus is on "doing things," not on the fact that it's "with words," and certainly not that grammar is involved, though in games like "What's that? Where's the x?" the motive of learning names for things is fairly close to the surface. There are "top–down" methods like this one for learning words, and there are "bottom–up" opportunities to learn words in games by inferring them from the context or observing the effects of their use. Parents gently pressure children to use more and more complex kinds of language by, for example, demanding explanations of what they want or are requesting, instead of mere directions ("Mommy open"). Parents ask for reasons, amplifications, clarifications, polite preliminaries, and graceful exits ("Say 'thank you'!" "Say 'bye bye'!"), usually without thinking about the language that's being learned in the process of learning to "behave" and "do things." But it is in this process of, to use Bruner's striking phrase, children negotiating their entry into a culture in conversation that many words are first learned.

Of course, first learning a word like *juice* or *electricity* isn't the end of the process of learning everything it means. A child's physicist mother may have a very different understanding of what electricity is than does the child to whom she's explaining the existence of sparks, associating the word, in Putnam's terms, with a different set of semantic markers or classifying features; a different "stereotype," or typical example; and a completely different extension in the world—but the acquisition of these initial toeholds tells us a lot about the things that happen later on, about what we should expect to happen in such subsequent occurrences as full-blown conversations: "This progress of similarity standards, in the course of each individual's maturing years, is a sort of recapitulation in the individual of the race's progress from muddy savagery" (Quine 1969:133).

How do games like peekaboo and Book Reading help children learn to interpret what people say to them? The things we say aren't meaningful without their full indexical contexts. Interpretation takes place in a whole world. But it's unreasonable to expect a baby to know very much about the indexical contexts in which utterances are being made or to be able to deal with contexts that are incredibly complex and constantly shifting. Because they're trying to make their way into our culture's complex shared world, it's not useful, when conversing with them, to assume that they've already done so. The advantage of a "format" like peekaboo is that it's a very simplified and repetitive context in which the basic tricks involved in interpreting utterances in a context can be introduced and practiced in gradually more complex and novel ways. This develops skills that eventually can be transferred into real, unplanned, unique situations.

Consider the various versions of the game played by Jonathan, one of Bruner's research subjects. Starting at three months and lasting until five months, Jonathan played a very rudimentary "pre-peekaboo" game, with all the initiative coming from his mother, who would sometimes cover her or his eyes with her hand and then surprise him by reemerging, perhaps (Bruner's account is silent on this detail) saying "Boo!" or "Peekaboo!" when she did so. This is typically how these games start: with one-sided participation by an adult directed at an infant who has no idea what to do, in which the child is introduced to the game by the method of assuming that he understands it, even though he clearly doesn't, and then gradually nudging him into fuller participation. From the beginning, the baby is treated as an interlocutor, with any crying often taken as an intentional attempt to communicate something, perhaps something physical like hunger, until about twenty-six weeks. The response is likely to be offering food or checking the diaper. After that, it begins to be interpreted as something psychological, with the interventions becoming increasingly things like offering a toy or attempting to engage the baby in conversation.

At five months, Jonathan's mother began to try something more complicated. She had a toy clown that could disappear into, and reappear from, a cloth cone mounted on a stick. Using this clown, Bruner divided her original game of peekaboo into an Antecedent Topic, involving Preparation (for example, "Who's this?") and Disappearance, and then a Subsequent Topic, involving Reappearance and Reestablishment.

After Preparation ("Who's this?"), the players decided who would be the agent and who would be the audience, though of course in the beginning, when he was only five months old, Jonathan could participate only as the audience, so this fork in the decision tree didn't become important until later. Then Disappearance, composed of three sequential components—Start ("Here he goes!"), Completion ("He's gone!"), and Search ("Where's he gone?")—was played. After a long pause, Reappearance could begin, either explosively or slowly ("He's coming! Boo! Here he is!"), followed by Reestablishment, made up of Arousal (moving the clown toward Jonathan, perhaps as if to tickle him?) and Constraint ("Oh no, don't eat him!")

Of course, every iteration of the game was different, with the stages being performed loudly or silently, quickly or slowly, creating a long experience of variations on the theme of a mysterious disappearance and a surprising return. Between five and nine months, as Jonathan became bored by some of the elements, they would disappear, since the two parties now could take them for granted. The crucial practice of omitting things with which an interlocutor is already familiar could be introduced at this time, and the mother could begin relying on an inventory of shared past experiences in conveying her message. As elements of the game began to disappear, the remaining elements were accompanied by more and more sophisticated uses of language ("Is he in there?" "Where's the clown?").

In the beginning, Jonathan's participation was limited to trying to grab the clown. By eight or nine months, however, he was reacting with smiling, laughter, and eye contact at appropriate points in the performance. Around eight months, he began attempting to intervene and to make the clown appear and disappear himself. At first his mother resisted, but when she did, he lost interest in the game, so she let him have more agency. Still, he soon became bored with the whole thing until the game reappeared in a more complex form, with his mother hiding a toy animal behind her back and surprising him with it. Jonathan developed a characteristic form of vocal participation at this point; he began to respond by making a raspberry. Later, his mother began to hide *herself* behind pieces of furniture. At twelve months, Jonathan suddenly and spontaneously began a role reversal, hiding behind furniture and reappearing with a cry of "Oo!" Finally, at one year and two months, the game with the clown reappeared—but with Jonathan and his mother now fluidly switching back and forth between roles. Sometimes he would use the clown to surprise *her*, accompanying the performance with attempts to replicate her vocalizations, with *a ga* for "all gone" and *pick* for "peekaboo."

Not only had Jonathan learned something about role reversal and managing shared attention, but his lexicon had been extended by the interaction as well. Since role-reversal imitation is something that chimpanzees almost never seem to do, it's interesting that humans possess these kinds of elaborate cultural technologies or innate adaptations for introducing children to the behavior. Learning to surprise someone is learning all sorts of things about attention and expectations.

The game quickly developed its own internal conventions, starting from natural meanings (the clown's disappearance was naturally surprising, and "boo" originally might have been a genuinely surprising sound for Jonathan to make when he reappeared) but gradually proceeding toward the more Gricean. As we play the game more and more, when someone says "Boo!" to us, it becomes clearer and clearer that he means to convey that we ought to act as if we are surprised and will be disappointed if we don't get the message that he's sending us that message. At some point, we discover that we can do the very same thing to convey to him that we mean to suggest that he ought to be surprised. Managing shared attention and meaning things nonnaturally aren't skills with which we humans are born in a fully developed form, as they might be if we were more like ants or crocodiles. These are things we must learn to do, and the process takes years. Without some sort of interaction with other human beings when we're the right age, we may not ever learn the skills in a perfect form.

Bruner uses Book Reading as an example of the development of reference. The game itself is simple. An adult and a child look through a picture book and attempt to converse about the pictures. Sometimes the child points at a picture, something in the room, or something somewhere else or tries to say the name of the things in the pictures or responds to questions about them. Sometimes the adult points to something in a picture and tries to elicit its name or a description.

In contrast to chimpanzees, who seem to have great difficulty learning to interpret indicative pointing (Tomasello 2008:38–41), human babies learn how to point, and understand pointing, fairly early in life, usually by around ten months of age, before they've acquired any real words (Tomasello, Carpenter, and Liszkowski 2007). This inherently Gricean and uniquely human (and cetacean [Herman et al. 2000]) communicative gesture begins as reaching, according to Bruner, which then becomes increasingly ritualized and eventually morphs into conventional pointing. Pointing then begins to be accompanied by baby words like *da* (that?) and then more specific elements of the infant's evolving idiolect, perhaps *apoo* for "apple" or *kiy-ka* for "cat." Pictures also evoke pointing. Both familiar kinds of objects in unfamiliar settings— a cup on someone's head—and unfamiliar kinds of objects in familiar settings—birds in the backyard—evoke pointing and may evoke idiolectic "names."

At this point, Bruner tells us, pointing and consistent naming (even if the names still are idiosyncratic to the individual infant) start to be used in games of "Where is the *x*?" and "What is that?" Reading picture books in a game that begins in this general context. Here's one of his sample early dialogues:

M: Look!
C: (Touches picture.)
M: What are those?
C: (Vocalizes babble string and smiles.)
M: Yes, they are rabbits.
C: (Vocalizes, smiles, and looks up at mother.)
M: (Laughs.) Yes, rabbit.
C: (Vocalizes, smiles.)
M: Yes. (Smiles.)

Later, though, the game becomes more sophisticated:

M: What's that? (Falling intonation.)
C: Fishy.
M: Yes, and what's he *doing*? (Rising intonation on final word, to indicate that this is the interesting and challenging part of the conversation.)

The mothers and children in Bruner's study started off without any Lewisian common knowledge of any kind. But as the game of peekaboo gradually coalesced and Gricean elements like role reversal and recursive mind-reading began to appear, knowledge that was common in Lewis's sense to both of them, things that both of them knew that both of them knew, began to appear. At first, it was unique to them: Who else (besides Bruner) knew that at a certain point in Richard's second year, *nini* temporarily meant "lady," having started out as a word for "juice," and *nani* temporarily meant "money"? But feedback from adults in games, and later in conversations, and the child's continued struggles to learn to be a party to his culture's various conventions gradually molded these local and temporary microconventions into more stable ones, which were much more similar to those of the rest of their speech community.

As Richard and his parents progressed from using ritualized versions of his idiosyncratic guesses or babbles to actual acts of participation by Richard in the community's prevailing linguistic conventions, his ability to indexically interpret utterances that depended on these conventions also was being developed. In the end, he must have become able to interpret remarks accurately even outside the repetitive and simple contexts of structured language games.

In response to the questions that began this part of the chapter, there *is* weeding and pruning, done mostly by the child, with a lot of assistance from adults or older children, whose participation in the process is indispensable. "Instruction and rehearsal" (Dennett 2009a) isn't confined to our acquisition of modern technical languages in the formal process of education. Plenty of "instruction and rehearsal" is necessary right from the beginning in the acquisition of any first language by any human, though the early versions may involve blocks, ducks, a cloth clown, a toy bow, or a picture book.

Once we know all this, we might wonder how late in life this process of establishing and refining or winnowing out microconventions, as well as learning how to interpret remarks that rely on the community's existing web of linguistic conventions, actually lasts. Is this exclusively a childhood activity; do we do this only in games like peekaboo and Book Reading? At some point, do we fully master the skill of interpreting the remarks of those around us and stop struggling to do it correctly, stop learning what particular words and particular ways of using them can and should mean? Are the language games of childhood the last games of interpretation that we ever play? Or does some similar, or functionally related, set of behaviors persist into maturity? Should we look exclusively to parent–child interactions as the locus for the generation and selection of new linguistic conventions and new interpretations of existing words, or does this process continue all throughout our lives?

If what I said in earlier chapters is correct, then what I said in this chapter about language games can't possibly be the whole story. I've compared words to sheepdogs, which suggests that the evolutionary fate of words isn't solely a consequence of their transmissibility to small children. They also must be serviceable for adult activities and must undergo further winnowing or modification during our performance of those activities. The dog must be friendly to children to survive as a domesticated animal, but it also must be very, very good at helping us herd sheep.

The basic model of the process that I've proposed is the sort of conversation that John and Twin John eventually might have had. That means it's quite important to understand what a conversation *is*, what it accomplishes, and how it's accomplished. It's to the phenomenon of conversation in general that I now will turn my attention, once again trying to analyze it as a game that involves challenging acts of interpretation. Now, though, since we're dealing with adults, or at least fluent speakers, these acts can become enormously more challenging . . .

9

What's Accomplished in Conversation?

GRICE'S THEORY OF CONVERSATION

What is a conversation?

This apparently simple question about an everyday activity we all participate in all the time is one of those unexpected puzzles about ordinary things that make philosophy so much fun. We think we know, surely we must know, it's when two or more people . . . you know . . . say things, and . . . exchange information . . . respond to each other . . . no, we really don't know. We have much better stories about lots of other apparently simple things we do, about why we eat or why we walk the way we do, even though those behaviors already are very, very complicated. But conversation seems harder to explain.

Why do we do it? Why do we tell one another these things? Why do we bother? What does it achieve? Why do we say the particular things we say, in the particular order we say them, in response to the particular things the other person says? Why do we laugh at particular points in some, but not all, conversations? Why is this one of our favorite things?

If it's a form of play, how is it related to chimpanzees' play activities? Are our jokes in conversation just more sophisticated versions of the gestural signals that chimpanzees send during play?

Our difficulty in accounting for conversation isn't a sign that nobody's ever tried to understand it. The intense focus on rhetoric by classical philosophers, for example, was the organized study of a certain rather formal kind of public conversation, and our interest in the phenomenon has continued until the present. Fortunately, I don't need to deal with all that history here. What I want to accomplish in this chapter is much simpler. Specifically, I'd like to tell you a little bit about H. P. Grice's (1975) model of conversation. (Because of everything else I've said, I'll also have to suggest a few extensions of the model.) Then I'd like to discuss what this might be able to tell us about the role of conversation as one place where nascent conventions may be born and where they're subjected to an analogue of artificial selection.

After talking about what allows conversation to serve as a tool for managing other kinds of culture, I'll finish by considering how these capabilities could have given rise to the very complex culture we live in today, which will finally allow me to try to provide an alternative to Daniel Dennett's (2009a) theory of the difference between ordinary and technical language.

In "Logic and Conversation" (1975), Grice argued that human conversations generally revolve around a presumption of a common purpose. We sometimes may be deluded in thinking that such a shared purpose exists—for example, when talking to a confidence man—but the supposition is required to make us willing to participate. The purpose may be obvious—the car is out of gas, we have to figure out what to do—or frivolous, extremely serious, or horrific. The torturer seeks to create a common interest so he can have a truthful conversation with us, even though his method involves the stick and not the carrot.

Grice admitted that he was perplexed about the exact nature of the understanding involved. Given what's been said so far, however, the puzzle doesn't seem insoluble. Common interests in the absence of enforceable contracts create coordination games, the kinds of games in which a Lewisian convention may be an equilibrium. Each of us would rather converse on some mutually agreed topic than not converse at all, provided that all the others do. It isn't a helpful form of participation in the conversation to periodically interject irrelevant remarks on completely unrelated topics, so we would prefer that all participants converse about the same topic as everyone else is or at least change the subject in culturally acceptable, legitimate ways. We often would be happy to converse about some other, slightly different topic if that topic had been raised by one of the participants instead. There always are alternatives, unless the people are enacting a play or another ritual, and real conversations change and drift as they go along, so the topic may well morph into one of those almost equally good alternatives. The conversation may acrimoniously disintegrate into no conversation, on no topic, if it goes badly, or it may gently evaporate into a resolve to have other conversations later. There always are different conversations we could have had instead. If someone new enters the discussion, we'd prefer that he stick to the topic, though if we were discussing something else, we'd prefer that he discuss that instead. A topic seems like a little temporary, local, mutable, Lewisian convention, or a set of conventions, established when we "convene" to discuss a certain subject.

At the same time, the topic is malleable, subject to the participants' direct manipulation in real time. A shared common ground is first established, and then it's extended and amended by the successive remarks of those involved. The changes may be incremental, or—if it's possible to bring the other participants along with us, if people are agreeable and the transition isn't too complicated to be made in unison without much preparation—they may be abrupt.

What's also true of most conversations is that not just anyone can participate. Perhaps we all haven't been properly introduced. Or the conversation may be one that only topologists or elk hunters or members of the president's national security team can engage in, or one that only Romeo and Juliet can be a part of. We may seek admission to a conversation and be welcomed or rebuffed. Yet this isn't usually because there's something scarce being shared by those conversing, which they would necessarily receive less of if someone else participated. Although there are conversations like that, many conversations are not. Sometimes new participants, even excluded ones, would have added something. In the language of economics, conversations are *excludable* and *non-rivalrous*. People can be prevented from benefiting from them or they can be excluded, but those who share in them don't necessarily diminish their worth for the others. Economists call such an association a "club."

It seems that a conversation—like the highway system or the community that speaks Welsh—is a particularly informal, spontaneous, and fleeting club, an ephemeral microinstitution that flickers into and out of existence in a few seconds, minutes, hours, days, weeks, or years after its initial convening and that is organized around a temporary set of conventions about its topic, manner, and so on. By seeking admission, we represent ourselves as willing to conform to these conventions unless we can persuade the other participants to amend them. Sometimes some of the conventions established in a conversation also acquire contractual force—for example, when the conversation itself is a negotiation—but many do not.

Note that Grice didn't say any of these things about a club or conventions. He was content to leave as a mystery the precise nature of the common project one joins by joining a conversation and to describe only the effects that the common interest produces.

How do the participants in a conversation pursue their common interest, whatever it may be? Grice asserts that much of what is conveyed in actual conversations between grown-ups is conveyed by what he calls *implicature*, a phenomenon that would be impossible without the presumption of shared interests. Consider the following exchange:

A: Will Susan be at the game?
B: She has to teach that day.

In Grice's terminology, B has implicated, but not said, that she won't be at the game. (To "implicate" is to create an implicature.) This conclusion depends on the common knowledge, known by both participants to be known by both participants, that teaching would preclude going to the game, perhaps because they will take place at the same time. Knowing this, A can work out what B is trying to tell him, what B is attempting to implicate.

Grice distinguishes between this sort of context-dependent, situational implicature, which he calls "conversational implicature," and mere conventional elisions of the following kind: "Socrates is a man, and therefore mortal." Here I've left out a premise that would be required for the "therefore"; I've neglected to mention that all men are mortal. But I didn't have to, because you and I, like everyone else, already know that. Without having to think about it, you naturally will assume that I am assuming that you will extract this information from the incomplete argument I've offered. Grice calls this slightly different phenomenon—in which Lewisian common knowledge takes the place of any working out of individual communicative intentions—"conventional implicature."

To work out the intended conversational implicature of an utterance, Grice thinks that we rely on various maxims, which he supposes are those necessary to pursue the presumed common interest behind any conversation effectively. He organizes these maxims under the more general principles of quantity, quality, relation, and manner. We assume that the speaker is telling us as much as we need to know for the purposes of the conversation, but no more (quantity). We assume that he's attempting to tell us only things that he knows to be true and is not asserting things that he believes to be false or for which he has no evidence (quality). We assume that what he's saying is somehow relevant to the mutually understood, though constantly evolving, topic of the conversation (relation). We assume that he's attempting to be perspicuous, that he would prefer to avoid ambiguity and obscurity, avoid prolixity, and present his narration or his argument in an orderly way (manner).

In answering A's question about Susan, B must be understood to be telling A as much as he needs to know for his question to be answered. Likewise, A must assume that B believes it to be true that she has to work and has reasonably good grounds for that belief. A must assume that this information is somehow relevant to the topic raised by his question. Assuming these things, A is in a position to interpret B's remark as intended to produce the implicature that Susan will not be at the game because it conflicts with her work. If her work has a special relationship to the game or its venue that means that the remark should produce the opposite conclusion, then B has failed to follow the principle of quantity correctly, because he's left out something he would have had to tell A to make his remark interpretable. He has assumed the existence of a piece of common ground that's actually missing.

In Michael Tomasello's (2008:57–59, 73–82) terminology, common ground is the common knowledge that participants in a communicative exchange have implicitly agreed is salient to their particular conversation, which is likely to include new items of common knowledge specific to the current interaction—for example, shared knowledge of the environment around them and their own recent history in it—as well as some old, standard items. Common knowledge, first created as common ground in formal or informal conversation and then conserved and referred to in later conversations, marks the boundaries of skill-centered speech communities, of the subcommunity of shamans or eel farmers or navigators or structural biochemists or Shinto priests. These are things that these people must know in order to converse with one another, making them unable to converse as freely with people who lack their skill set.

In conversation, the method used for creating new items of common knowledge is the participants explicitly or implicitly informing one another of things, so conversations create parts of their own common ground as they go along. By so doing, the participants may become partly isolated from the rest of their speech community, which now doesn't share the newly created common ground. New tacit conventions also are negotiated indirectly and obliquely in particular conversations, by means of concerted choices among competing, unstated alternatives, which can make it even harder for an outsider to follow them.

A conversation consists of a series of its participants' dovetailed and cumulative modifications of their common ground, and at the end of the conversation, they may share different knowledge or intentions ("Then yes, let's do that") or expectations ("Well, then I guess we can expect the same thing to happen every time") or different explicit conventions ("OK, I guess next time, whoever called originally should be the one to call back") than they did before.

This new knowledge has become common knowledge in the group conversing and now can be used as such, can be assumed to be part of the common ground for subsequent discussions by the same group. B will expect A to remember that Susan has to work on the day of the game. We'll be expected to remember the new plan or expectation or convention that's finally been arrived at.

Although this wasn't part of Grice's story, one of the things that happens in a conversation is that knowledge, expectations, or beliefs are converted from private knowledge into common knowledge within the group conversing. What is common knowledge can support conventional (as opposed to conversational) implicatures, so the group's stock of possible conventional implicatures is enlarged as a result. From now on, it may not be necessary to mention that Susan has to work on the day of the game; perhaps it can simply be assumed. Every successive conversation among a certain group means that less must be said in subsequent conversations, that more and more can be "taken for granted."

This fact can produce a sort of cultural microversion of songbirds' local dialects, local, group-specific assumptions that make it harder and harder for newcomers who lack the same shared history to participate in the group's conversations. Conversations make us clannish; they erode the barriers to communication and trust *within* the group while erecting new ones *around* it, in a tiny, temporary, ultrafast cultural version of one of John Maynard Smith and Eörs Szathmáry's (1998) "major transitions." A conversation creates a club that subsequently may function in some ways like a single, self-interested unit, which may see itself as competing with other, rival clubs and may exclude interlopers or impose its own rules on new entrants.

Grice points out that many of the same considerations about presumed common interests, and the need to seek admission to the cooperating group, are active, not merely in any conversation, but in any coordinated and dovetailed interaction among people.

For example, I've run out of gas in the desert. A passerby stops to help me. He hesitates until I acknowledge him. He raises an eyebrow, which I interpret, on the basis of principles about relevance and perspicuity, as an instruction to share with him the nature of my problem. I wordlessly point to the cap of the gas tank. He smiles and fetches the gas can from the back of his truck. If he'd smiled in the same way, gone back to his truck, and simply driven away, I might have been annoyed or at least surprised. Why stop, inquire about the nature of the difficulty, act as if you intend to be helpful, and then leave?

The passerby sought admission to the club of persons grouped around the problem of my stopped car and working toward the common purpose of getting it running. I might have been alone in that club, but he still had to seek permission to enter it. Then, on the basis of the details of the situation and universal common knowledge, he sent me an easily interpretable conventional signal, raising his eyebrow, to which I replied with a gesture that made sense only in context, pointing to the gas cap—and yet now he's walking away without doing anything about the problem that became partly his when he sought admission to the club. This is not the convention for how we behave in this situation, it's a violation of obligations that were assumed voluntarily, and I may frown at him or apply some other mild sanction (mutter "Well, thanks for nothing") in a halfhearted attempt to police the borders of the club against such inconsiderate incursions.

Thus the class of human situations that Grice calls "conversations" is much wider than that term seems to imply. No actual language has to be involved at all. Any situation in which tools are used or made cooperatively, or one party assists another in some task, or one party shows another how to do something, or what something is, or in which people share food or drink tea together already has these features. One must be admitted into the group of people doing the work, and on the basis of assumptions about their contribution to a common purpose, one must interpret gestures and actions as reflecting particular intentions in a way dependent on the suppositions that they tell us what we need to know and not more or less, that they're not intended to be deceptive, that they're relevant to the task at hand and aren't hopelessly obscure, and that the gesturer has come up with an instruction that we should be able to interpret.

When the master holds out his hand for a hammer, the apprentice can understand the gesture as a request of that kind only because he assumes that the master isn't making an unnecessary gesture, isn't trying to trick him, is asking for something relevant to the collaborative task at hand and not his hunting spear, and isn't making a gesture he thinks the apprentice will be unable to interpret. It's these things that no wild chimpanzee would ever know to assume, but they're essential to both linguistic and nonlinguistic communication in humans. The same assumptions are central to the special kind of conversation we call teaching and learning.

Although human children learn to assume these things, "getting" remarks made by adults in conversation can be quite challenging. If they're not sharing attention with the adults, they may not even try: the adult is likely to seem to them like one of the adults in cartoons who move their mouths but can make only meaningless noises come out. Participating in the conversation of any particular club is a skill that must be learned and may require specialized knowledge. Since it determines to a considerable extent whether we'll be accepted as a full member of the club, it's an important skill. Indeed, in the long history of the human species, it has often become a matter of life and death.

A rather difficult thing for an adult to learn and remember is what is *not* common knowledge in a club, what you would think anyone would know that these people don't know or don't believe. If your remarks assume knowledge that other people lack, if they presuppose a degree of sophistication that isn't present, the remarks will seem outré or stupid to the members of the club, which may partly explain Charles Baudelaire's ([1861] 1974) comparison of the poet and the albatross:

Exiled on earth amongst the shouting people,
his giant's wings hinder him from walking.

An activity of creating and maintaining a shared picture of the world and the possible worlds around it is basic to any human cooperative task. The activity of conversation seems to have the dual purposes of simultaneously extending this shared picture of the world in various directions and making sure that it *is* shared, that conceptions of the cooperative task and its circumstances aren't diverging too rapidly as it's performed. If they did, the dovetailed nature of the work could easily lead to disaster: You might think I'm planning to gather firewood, and I may think you're planning to do so, and the fire may go out one night. Or you may think I'm watching the children manage it, and I may think you are, leading to the death of both of us in a forest fire. To prevent such catastrophes, we need to talk to each other frequently and thrash out our misunderstandings. Language must have evolved at least partly to serve this shared project, the creation and management of common ground, of a shared, reasonably accurate, and up-to-date consensual picture of reality, which seems ultimately to derive from the cooperative employment of tools or more complex items of culture like fire making.

Cooperation is something chimpanzees may engage in a little. They hunt and patrol together; bonobos seem to cooperate in enjoying sex. They also use tools. But in the wild, they never use tools cooperatively and do so in captivity only if they absolutely have to. Tomasello (2008:176) pointed out that you will never see two chimpanzees cooperatively carrying something. How could they? They have no way to learn how to work on recursive projects with deferred goals, so in carrying something together, they'd each be seeking a tantalizing immediate reward. If each of them actually wanted the thing they were carrying, each of them would also want to deny it to the other, so the attempt to cooperate would quickly become a struggle. The motivation required for success—the idea of working together toward a shared, distant goal—is inconceivable to an unhabituated chimpanzee.

It could conceivably have been something as simple as cooperative carrying—carrying a scavenged carcass back to the trees together before the lion arrived or carrying firewood together—that began the human-evolution ball rolling. Such an activity would make it natural for individuals to take an interest in the way that the other individuals performed their roles, and for skilled individuals to begin to weed out inept performances by the others. In this situation, a blow to the head conceivably could shift its meaning from "I'm going to kill you if you don't get out of here right now" to "I want you to understand that I want you to understand that you need to pull your own weight." As Kim Sterelny (2012) suggested, whatever the cooperative activity was, it may have served as an early form of teaching, with the inexperienced individuals assisting the more experienced ones.

Once this technology for engaging in culturally transmitted versions of dovetailed, coordinated cooperative activities got started, it must have become more and more elaborate until we arrived at the tremendously complex cooperative activities that characterize modern human societies. Managing such a complex culture of cooperation by using a complex modern human language, however, apparently requires constant, very complicated efforts at disambiguation.

DISAMBIGUATION AND HUMOR

Disambiguating a remark means choosing one interpretation over another, so in a process of semantic evolution, disambiguation is a form of selection acting on competing interpretations. His account of disambiguation is the most interesting part of Grice's (1975) story about conversation.

I earlier introduced some of this account: B's reply to A's question about Susan is potentially ambiguous. What does that have to do with anything? Does it mean that she can go or that she can't go? What exactly is B trying to implicate? If he has the requisite information, A will be able to work out the intended implicature by assuming that B is attempting to follow the relevant principles, that he has additional knowledge that he hasn't mentioned, and that he has reasonably assumed that from the context, A will be able to work out what it is. To understand an intended implicature is to disambiguate an ambiguous or incomplete statement in this way.

Sometimes, however, there's more to the problem than just missing information. Suppose you ask me where Susan is now, and my answer is, "She's somewhere in the Western Hemisphere." Well, that's not a very useful answer, since it won't really help me find her. Couldn't I have told you where in the Western Hemisphere she is? I've violated the axiom of quantity; I haven't told you enough. But why?

There might be various reasons. Perhaps I've come into conflict with the axiom of quality: I could make up a more precise answer, but I couldn't be sure it's true, and I prefer not to risk misleading you. But other explanations are possible as well. Perhaps, given your history of going out drinking with Susan and starting fistfights in bars together, I'm simply unwilling to answer and am opting out of that part of the conversation by refusing to continue it in accordance with the principles. In each case, a different implicature must be worked out.

Sometimes, in order to deceive, people also quietly, unobtrusively, violate the principles.

The most interesting thing I could be doing, though, is openly flouting one or more of the principles. Notice the intrinsically higher-order character of "flouting." Not only can you know that I'm violating a principle, not only can I know that you know, but you also can know that I know that you know, and I, knowing all that, can insouciantly act as if I don't care. That's what flouting is, so we're on ground that should be rather familiar by now, dealing again with recursive third- and fourth-order expectations.

I'm deliberately violating a principle, and I must know that you know I'm doing it, so when I openly flout a principle in this way, I don't have the goal of actually deceiving you. I actually want you to know that since the superficial meaning of my words represents a violation of the rules of conversation, I must mean something entirely different by them. I'm making what we call a "joke," displaying "wit," as we humans do. Perhaps Susan is a colleague who never is anywhere to be found when there's work to be done, and what I'm really trying to say is "Missing, as usual." Or perhaps I know that you need to find her right away, and my intent is to hint playfully that I require persuasion in order to divulge her whereabouts.

In all these cases, I seem superficially to be violating the principle of manner because my way of conveying the information seems unnecessarily obscure. An example Grice gives is "You're the cream in my coffee." Since you are not literally cream, this is, superficially at least, also a violation of the principle of quality, in particular the maxim that we should avoid saying things that are untrue. However, it makes sense to suppose that what the speaker means is that you are a source of pleasure to him, like cream. But if he says it in response to your wrecking his car, this may not be the preferred interpretation after all. You may be being asked to make a double leap, involving two successive inferences. He may be suggesting, first, that you give him pleasure and delight, like cream, and second, no, not really, he means the exact opposite.

Why not just say that? Why not just say "You're a nuisance"? Why deliberately make the remark in a manner that strains our interpretive capabilities to their limit, requiring us—on the basis of subtle cues, obscure bits of background knowledge, or quirky features of the common ground established in the conversation so far—to suppose that we're somehow like cream but really are the opposite of whatever way we might be that would make us similar to cream? Why force us to go through all these mental gymnastics just to disambiguate a casual remark? Grice's analysis stopped with the idea that humorous remarks propose these sorts of interpretive puzzles; he didn't speculate about why human behavior has this interesting feature. Given what's been said so far, though, we might be able to make a few guesses.

In a biological and evolutionary context, the obvious answer is "because they're gymnastics." Attaching shared interpretations to communicative acts and coordinating our behavior around those shared interpretations is something that humans do. It can be difficult, and we often fail. Children often may be unable to do it. They have to learn how, and it takes a long time to learn an adult version of the skill. You often have to pay close attention to do it, and you need to remember what's been said so far. When you make or get a joke, you aren't just complying with the social contract, you're doing so in an exemplary manner.

Humor is a form of *play* that exercises the skill of interpreting, of eliminating ambiguities and figuring out what's really being said. This associates it very closely with the evolutionary roots of human language, which, as we've seen, appear to have some connection with the sorts of gestural signals that chimpanzees use mostly in relatively inconsequential situations like play. But the chimpanzee's form of play seems much easier than what we do. It doesn't seem to require the tremendous skill that people use in playing games or dancing or making art or making other people laugh. Getting jokes, in particular, can be hard, as hard as solving a chess problem. Even adults don't always succeed in correctly interpreting these playful remarks, and people differ markedly in their ability to propose and decipher such little riddles. Some people are funny, and others are not.

Just as birds often court their mates with aerobatic displays, because flying is something birds can do, have to do a lot, and may differ in their ability to do, when people make humorous remarks, they may well be engaging in what is, from their individual point of view, a sort of acrobatic display of mental strength and agility. At the same time, they're also suggesting a point of view to the group. The remarks are supposed to be difficult to understand, but when you do finally get them, you're supposed to be persuaded by them. "Yes, Susan often is very hard to find, isn't she." "Yes, it is a bit careless to wreck someone's car, though perhaps since the driver is still being asked by the owner to play a pleasurable and light-hearted puzzle-solving game, indicating that both are still in the conversational club together, it isn't unforgivable." The chain of inferences required to interpret the remark constitutes a sort of argument by itself, with the psychological reward of laughter at the end as a prize for following it all the way to the intended implicature. Few forms of argument are more persuasive.

The conversation of a person whose remarks pose no such interpretive puzzles or always point toward interpretations with which we're strongly disinclined to agree or don't seem to lead anywhere, can, under the wrong circumstances, become almost unbearably tedious or irritating to us if the interaction lasts long enough. Humorous remarks are supposed to incorporate a point of view that's a little difficult to arrive at but convincing once you get there. It's a point of view that isn't just the invidious, narcissistic, stupid, hypocritical, or disgusting fantasy or misinterpretation of the joker, at least as far as the conversing group is concerned. (Of course, depending on its composition, the group's standards for these things may be arbitrarily low.) Failure to conform to these expectations may cause the group to laugh "at" you and not "with" you, to loudly arrive together at an interpretation very different from the one you've tried to lead them to. (Perhaps it was *you* who wasn't paying close enough attention.)

Laughing when someone falls down or has a pie thrown in his face, I believe, is simply a variation on the same theme of "laughing at," a shared recognition that the person has inadvertently and virtually declared himself to be clumsy, incompetent, easily distracted, or unreliable while trying to accomplish something else, or else he has very cleverly mimicked such an inadvertent declaration. Consider the Reverend Spooner's assertion that the Lord is a shoving leopard. The image is incongruous, but it seems to me that it is the possibility that the reverend may inadvertently have succeeded in telling us something meaningful about the contents or character of his faith—and the strangeness and triviality of what that meaningful something would have to be—that makes it funny. The malapropism wouldn't be as amusing if it didn't come from a reverend speaking about the Lord.

Being laughed at can be an unhappy experience for a creature whose fitness depends on its ability to ostentatiously cultivate flattering forms of self-presentation.

Of course, in our judgments of others, it's better to be more charitable than people often are. Laughing at a mimicry of ineptitude, unreliability, inattentiveness, selfishness, unhelpfulness, or stupidity is sometimes completely harmless, though, because the mimic may be taking an impartial, or even exaggeratedly critical, view of her own foibles or those of some would-be authority.

The notion of "wit" thus is a somewhat oxymoronic combination of interpretive trickiness and goodwill, impartiality, the will to see things from the perspective of the whole group or a typical member. The joker is apparently being unreasonable. She may seem to be denying analytic truths or may speak of pigs flying and streets being paved with gold, but if she's done her job right, in the final analysis it will become clear that she isn't being unreasonable at all, that it's a fair comment. What could be better in a human than to have *both* trickiness and goodwill at the same time? What could be less impressive in a human than the simultaneous combination of poor judgment or gullibility and ill will (revealed, say, in an attempt at a caustic and ungenerous remark that backfires badly)?

Successfully making a joke in conversation isn't just a display of culture, like playing a piano sonata or dancing a pavane. The joker is deftly managing clashing conventions within a culture, coming up with ways of apparently flouting them that actually fulfill them. Not only does she *have* culture, but she knows how to make culture work well under challenging circumstances, how to fit it into the complex real world, how to interpret it in a useful way, how to use her brain for what it's for, how to deal with the unique and with imperfect analogies, even if this sometimes requires making a new piece of culture or modifying an old one. Presumably, this attracts us to the person and produces a happy form of arousal in the group, partly because it's very good news about a human or a group of humans, in Darwinian terms, that they're this good at what's more or less the central human adaptation.

These laughing people know how to share and manage a consensus reality that isn't irreversibly drifting out of alignment with the world, that isn't so rigid or hypocritical or narcissistic that it will kill them. The pleasure that accompanies laughter is a pleasure in succeeding in a difficult, shared interpretive task that can counterbalance the older pleasures of food and sex. It's a pleasure that can make creatures like us devote time and attention to the endless and often somewhat challenging task of keeping up with the tacit and constantly changing conventions of the group's consensus reality. It's originally and essentially a shared pleasure because conversation and interpretation originally and essentially happen in social groups, even if humans now can have conversations with themselves and perhaps even laugh at their own jokes. (That's still much harder for us than laughing at someone else's joke.) It's adaptive for us to be pleased by everyone's arriving at the same complex and unexpected but not unreasonable interpretation of a remark, because this is what we have to do in many other, more directly consequential cooperative behaviors.

A chimpanzee can be playful, but as far as I can tell, it lacks this sort of sense of humor. It seems to me that joy in discovery as such, which is a related phenomenon, may be found only in creatures with a complex language and a complex culture, which constantly poses this sort of interpretive problem. It's an adaptation, one that helps us all keep up. A chimpanzee probably feels joy at finding a nut, but I'm not sure he feels joy in discovery.

But we're dealing with humans and human culture, so things are more complex than that. The behavior of birds when they sing a learned song may offer another lesson here. Although an older bird's song, from his own selfish point of view, is simply a display, it also serves as a model for younger birds who are still learning, helping homogenize and render distinct the local dialect. Similarly, humor may have another function. It may serve as both a display, by the joker, and a reward for the audience member for keeping up, in the way I've described, and also as a way of homogenizing the local dialect over the longer term, as a way of fixing and filtering shared interpretations.

These little exercises in acrobatic forms of disambiguation seem like ways of keeping synchronized everyone's picture of what's going on, by constantly testing whether we can anticipate one another's unstated thoughts. When we can't, we often supply explanations until the others laugh, having finally succeeded in following the interpretive trail. Our shared system of imagined possible worlds is brought back into harmony in this way, even at points rather distant from the actual world. In the absence of some such process, our individually varying idiolects might drift too far apart, impeding communication. Our ability to coordinate is being strengthened by the process of disambiguation, which strengthens the social contract. All our idiolects have been strained through the same fine interpretive filter. Descriptions and interpretations have been supplied where necessary. The joker, acting as a *nomothete*, has succeeded in carrying us all along with him as he comes down in favor of a particular complex interpretation of his words and the situation, out of a sea of possible rivals.

What does it mean to interpret a situation, as opposed to a sentence? Presumably, it is picking out which set of possible worlds is like the actual world in which the situation exists, which means deciding which features of the actual situation are the preeminently salient ones, the ones that ought to be obvious, the ones that imply "natural" analogies with other possible situations. It's to propose a particular judgment about what's important about the situation as the most natural one. So in "interpreting a situation" by making a humorous remark, we seem to be trying to control which worlds will be seen as being nearby. With the remark about Susan, the speaker seems to be suggesting that worlds in which she is nowhere to be found tend to be quite common in the region around the real one. We're managing and influencing the metric that will be used to measure what's an outlandish interpretation of some subsequent remark and what's a perfectly reasonable one.

We enjoy trying to achieve this kind of harmonization, so if we can, we'll continue doing this until we get it. By the end of the process, many things that are implicit, or tacit, conventions or expectations or implicatures or any such thing will have become explicit. Humor, it seems to me, is a locus of selection for meanings, for shared interpretations, bringing back together the complex associations and difficult-to-identify referents we collectively attach to words and phrases in those places where they might easily pull apart. The activity's acrobatic character makes the mesh of the filter finer, makes it harder to get the joke, and makes it more imperative that we understand its constituent words and every overtone of their meanings or possible associations in exactly same way the joker did.

We may resist this challenge, balking at the fence, refusing to leap. ("No, I don't think that's funny at all." Or in other words, "The trail you've laid down is misleading, you jerk, those aren't our conventions, you're not giving words their obligatory associations, there's nothing even remotely analytic about the statement you just made. As I understand the words of our shared language, that isn't what's important about this situation at all.") The proposed interpretation may even make us very angry, but when we finally laugh, we've clearly surrendered, we've admitted that the emperor is naked, even if we don't like admitting that. If we laugh, the joker isn't being unreasonable. If we laugh, her associations are obligatory in the language of the group. The features of the situation that she thinks are important really are important. If we laugh, what she's said may really be analytic, may really follow from what are now the ineradicable connotations of her words, which is why we often try so hard to resist laughing.

In laughing, we haven't just lost control of the conversation; we've lost control of the common language and the consensus reality that it's suited to describe. We've lost control of the whole system of possible worlds, of the features of each world that should be considered salient, and of the metric of similarity that should be imposed on them. Now what the joker thinks is obvious really is obvious to everyone—but obviousness has the kind of tyrannical power in human affairs, as I explained in chapter 2, that can make this a terrible and damaging concession.

This probably is why political satire is so frequently suppressed: the tyrant genuinely can't afford to look obviously contemptible; he can't let his actions be subject to public interpretation by someone else. "They pretend to pay us, and we pretend to work" is a complete, unrecoverable political disaster for you as a public interpretation of your actions if you're the one who's now understood to be only pretending to pay people, whose actions are being successfully named a "pretense" and compared with other silly pretenses, such as the use of toy money in Monopoly.

The joke-in-conversation is a site of selection for interpretations of conventions, such as our convention about what actually counts as "paying someone" or what actually counts as "working," a place where conventions can be sharpened, renegotiated, reinterpreted, applied to novel objects, or even born, as an implicit logic of a situation made explicit through a difficult but persuasive new interpretation. The human community that laughs together also interprets together, and when we all interpret in exactly same way, we share exactly the same (possibly amended) conventions regarding meanings.

This amendment in our interpretations may be a source of regret for some of us, who might wish that words meant something different from what they now apparently do in this indexical context. But once we, or the others, laugh, this fight already has been lost. If we can, it is better by far to join in the laughter, however damaging it is to our interests, than to be left on the sidelines speaking a slightly different language from the rest of the group, in constant danger of being seen as a liar and an enemy of the social contract or else simply stupid, witless, if we persist in pushing our discredited interpretation, continue to insist that the useless play money we're all being paid with is real money.

The human who routinely fails to follow these interpretive obstacle courses runs the risk of himself becoming the target of the jokes made by the group. They may become complex and acrobatic ways of disparaging *him*. But laying down trails that nobody else could or would follow because the knowledge presumed to be common is not, or some other inference is more obvious, or the trail leads through unpleasant and disgusting places, is also likely to be perceived negatively: "His jokes aren't funny. They're either stupid, or gross, or they make no sense at all."

We must carefully steer between the banal, on the one hand—remarks that are too expected and too easy to interpret, which make us seem dull, slow, and timid, or remarks that accomplish nothing new with respect to the group's interpretation of its shared language or its shared picture of the various nearby possible worlds—and the obscene or the obscure, on the other hand—remarks that make us all think about the fact that we're all thinking about something we don't want to think about the others thinking of us as thinking about, that make us wish we could put the fig leaf back, or that don't make the members of the group think anything in unison at all. These sour notes, remarks that bring us to no common destination or bring us all to a place we'd really rather not be, occur in many, perhaps all, conversations, but unless someone wants to follow one up for some reason, they're often simply abandoned as dead ends. The failed joke is never a political victory and doesn't do much to twist the group's language in the joker's favor.

Supplementary witticisms may help us all get on the same track by the time we're finished exploring the whole joke, once all the further outré and just barely interpretable or decent remarks about sour cream and Irish coffee and wrecked cars that might occur to anyone have been made and collectively and simultaneously chuckled at. These follow-up jokes seem to have the function of making sure that we all now see and expect others to see and expect others to expect us to see the same system of possible worlds, with the same metric of similarity imposed on it. Some very strange and distant worlds are mentioned in jokes. We must see the same ones at the same relative distances in order to laugh at the same things at the same time, and that involves harmonizing our notions of which ones are closer or farther away, adjusting our idiolects so that even in the far-off worlds to which David Lewis ([1969] 2002) thought their clashes were confined, they instead match up.

When we all have laughed at the joke, everybody knows that we all know that we all are thinking the same thing, so a new piece of common knowledge has been established in the group. It subsequently can be referred to, and people may be expected to remember it.

Laughter itself is an extremely interesting phenomenon. It's an involuntary, reflexive vocalization, like scream of pain or a chimpanzee's pant hoot of cheerful excitement. It's specific to humans and very difficult to "fake," to induce voluntarily, in a convincing manner. Just as telling a joke is an unfakable signal of interpretive acuity and in-group membership, so is sincerely laughing at a joke. Laughter in chorus, in a group, is a sound of pleasure about having arrived by a difficult path at concordant interpretations.

No doubt Dr. Tulp accompanied his anatomical lecture with a number of humorous and self-deprecating remarks, or anyway, many of us would have. Once your audience laughs, you have them. After that, they're trying to "get" the new interpretation or new information you're offering them, not looking around the room for food or sex, as a chimpanzee would. There's this other pleasure with which they can be distracted to stop them from thinking about such things, which merges seamlessly into the pleasure of interpreting the talk, because it, too, is a pleasure of interpretation.

The individual who doesn't really get a joke can pretend that he did, but the laugh is likely to sound fake, since it's a voluntary imitation of an involuntary response. The faker will be easy to trip up, for example, when a humorous remark is made in a deadpan way. (We'll do almost anything to make interpretation hard, but not too hard.) He can't always get to the right interpretation fast enough, but because he is bluffing, he eventually will laugh in the wrong place or at the wrong time all by himself, or he will say something that members of the club think is incredibly stupid, so this is ultimately a self destructive choice. The signal of laughter is a boast that can safely be made only when it's true, a claim of strength, of an ability to participate in a cooperative enterprise in an alert and aware manner, to interpret even difficult remarks in unison with the group.

If you really don't get it—which is everyone's situation immediately after entering an existing skill-centered group and may even be a sign that you're just not ignorant enough to be one of these people—but for some reason you want to belong, you're better off making yourself useful in a subsidiary role while you try to figure out what everyone here already knows (or doesn't know). To have difficulty faking laughter is to be safe from the self-destructive temptation to overpromise in this regard. Our irritation at people who merely pretend to get jokes seems like an evolved response because people are so prone to respond in that way, so utterly unable to resist the impulse to be cruel to the faker.

The joke and the answering group laugh strike me as the human answer to the hummingbird's hovering courtship dance or the nightingale's hundreds of dangerous and complex songs. We must make these spectacular exhibitions of interpretive puzzle making and interpretive puzzle solving at least partly to show that we can. But doing this repeatedly also helps us nudge back into alignment each person's idiosyncratic picture of the common ground of interpretations established so far in the conversation, so this exuberant display has another function. It's also a mutation-repair system, a form of purifying selection, or a filter that admits particularly apt innovations into our language and its interpretation, nudging all our expectations back into line with one another and making them more subtle. The joker is a kind of nomothete, and all of us who do or don't laugh at her joke are playing the role of Plato's dialectician or teacher, the role of Socrates, deciding which of her proposed interpretations we will accept, which of the slightly modified sheepdogs we are or are not willing to buy. The evolution of our language is ultimately driven by our individual choices about what words to use and what sense to use them in. But those choices don't happen in a vacuum; they happen in conversation, in the course of our efforts to accomplish what's accomplished in conversation.

ADAPTIVE FILTRATION AND ADAPTIVE IMMUNITY

This sort of intraconversational selection happens very quickly. Ideas come up and die within seconds; comments are made on comments on comments; conversational objections, or mere failures to find something amusing, can move equally quickly. Cultural evolution is much slower, as it takes time for a convention or a skill or a verbal mannerism or a hairstyle to percolate through a population. Biological evolution of the underlying cognitive machinery is slower still. How are these three different processes supposed to be related?

The organism evolves slowly, which would make it helpless against the possible maladaptive tendencies of the more rapidly evolving culture unless it had an even faster countervailing process, conversation, with which to control it. In this sense, words are like agile, obedient sheepdogs, and other cultural traits are like slow, balky sheep. I argued in chapter 8 that we learn language while negotiating our way into a culture, but its actual employment by adults often involves an attempt to negotiate their way *out* of some trap that their cultural inheritance has inadvertently set for them. (Often that's what comedies are about. Don Quixote interprets too literally the traditions governing his role as a knight, and Sancho Panza's way of dealing with the nobility is unfortunate.)

A similar set of intersecting selection processes can be found in another domain, in the three-cornered coevolutionary (or quasi-coevolutionary) interaction among populations of organisms, their pathogens or parasites, and their immune systems.

The organism must deal with an inherently unpredictable challenge from microbial pathogens that, because of their shorter life cycle and larger numbers, evolve more rapidly than we do. The difficulty is exacerbated by the fact that its own attempts at defense are the most important selective factor in the pathogens' evolution. Whatever the organism comes up with, they'll be selected to work around it. The organism is caught in an evolutionary arms race with its targets, and it has to keep changing its response to stay one step ahead of them.

The organism manages to do this partly by creating a miniature, controlled evolutionary process within itself. A set of sequences on the genomes of leukocytes, a population of cells in the human immune system, are repeatedly scrambled together into a very large number of combinations. Each clonal line of leukocytes produces a different scrambled-up sequence. These different permutations or combinations are then transcribed into RNA and translated into chains of amino acids.

This system of scrambling and mutating sequences—V(J)D recombination and all its various sequels and accompaniments—has some of the same complexity as the syntax of a human language, which also is a system for permuting strings with strings and selectively altering the results (Jerne 1984).

These chains of amino acids fold up into protein molecules, "antibodies." Because the genomic sequences from which they were transcribed and translated were the result of scrambling a set of elements into any of a vast number of possible permutations, the chains thus generated fold up into a vast number of different shapes. By chance, some of the shapes will be complementary to small shapes found on molecules that are on the outer surface of a potential invader, a parasite or disease-causing organism. When an antibody finds a target to which its randomly generated surface structure can bind, the clonal line of leukocytes that created it begins to multiply rapidly, and as long as the antibody is still finding antigen to bind to, the explosion in its population will continue. The molecules with antibodies tagging them and the intruding cells or viruses to which they belong are then destroyed by other kinds of immune cells, which attack anything with such a tag on it.

What did our own ancestors need a new kind of defense against? If the arguments given here are correct, they needed a defense against noise and maladaptive culture. Achieving a plateau in the amount of culture that can be conserved, as the inventors of the Acheulean technology may have done, since their technology was preserved apparently almost unchanged for hundreds of thousands of years at a time, doesn't mean that there's no further scope for invention and critical judgment. Because useful ideas are constantly corrupted or lost, they should have had relatively short half-lives, even if there were far fewer of them. As we learned from Magnus Enquist and Stefano Ghirlanda's (2007; Enquist, Erikson, and Ghirlanda 2007) models, they must have been continually reinvented, and the corrupted versions must have been repeatedly recognized as such and weeded out.

The users of that early technology were probably much worse at weeding out maladaptive practices than we are, if ψ is something that can evolve over time. They might have been carrying a considerable burden of maladaptive culture. They presumably would have been locked in a long coevolutionary arms race with it, a race that our ancestors later won when ψ became high enough to allow the accumulation of unlimited amounts of culture. But until our adaptive filtration finally became so good that there was no upper limit on the amount of culture we could usefully maintain and we began to become fully modern humans, the limiting factor would always have been the accumulation of maladaptive culture through the corruption of originally useful practices or the invention of harmful new ones.

This difficulty is similar to the one the immune system faces. Our defenses against maladaptive culture must be the main selective force acting on that culture. Whatever we were capable of doing, our maladaptive culture would have evolved a way of countering. We would have been manipulated and deceived in ways that were essentially reflections or negative images of our own adaptive-filtration capabilities, what they had achieved by repeatedly and over a long time selecting out all the less able deceivers.

The altitude of your plateau, the amount of culture you can successfully maintain without drowning in your own mistakes, depends on how efficiently you can filter and reinvent it. From the fossil record, we seem to have had a long period of stability with a much smaller amount of culture, and probably a much simpler system of communication or language, than modern humans have. Apparently these tasks are quite difficult for creatures like ourselves.

But in the past several hundred thousand years, this has changed again, first with somewhat more modern-looking people, and later, starting only a few hundred thousand years ago, with a more complex material technology that finally broke from the Acheulean mold and the beginning of an unlimited exponential process of endlessly acquiring new items of culture. The amount of material culture maintained by human societies, for example, the number of kinds of items made, began a gradually accelerating expansion at that point, which has continued at an ever more rapid pace until the present. Somehow we overcame the difficulties associated with creating and maintaining a more complex culture that had kept us stuck at a simpler level for so long.

This is the analogy I want to convey: both we and the vertebrate immune system faced a similar problem of a rapidly evolving challenge. Our ancestors faced the challenge of maladaptive, parasitic, obsolete, or corrupted culture and the need to constantly weed it out and invent replacements in order to be able to maintain a larger amount of adaptive culture. Vertebrates faced a challenge from actual parasites and microbial pathogens. Both evolving lineages eventually arrived at a similar solution, a form of rapid simulated counterevolution built around the generation of variation in real time by a context-sensitive generative grammar.

The generative grammar for antibodies creates antibodies whose structure may be complementary to the structures of aspects of the problem for which they exist to deal with, of aspects of molecules on the surfaces of pathogens. Our generative grammar for sentences, for imperatives and descriptions and questions and so on, may have as one of its functions generating sentences whose "structures" in some broad sense are complementary to the structures of aspects of *our* problem, allowing us to discourage, forbid, describe, critique, praise, mock, or otherwise converse about billions of different, specific, potentially maladaptive, parasitic, obsolete, or corrupted cultural behaviors (as well as the occasional praiseworthy innovation). In this way of looking at things, "Thou shalt not kill" is functionally like an antibody against casual murder.

When behavior can be described, it can be sorted through and evaluated, like a group of dogs at a dog show or pedigreed pigs at a state fair. One end that the description of human behavior by humans can serve is to advance the domestication of human culture. Language allows us to discuss an infinite number of different describable behaviors and deliberately weed out the ones that don't please us and explicitly propagate the ones that do in a way that would be impossible if our powers of description were more limited. To help us manage our domesticated culture, language itself had to become domesticated, so language also is used to manage language, the way that Lewis used it in writing *Convention* ([1969] 2002), the way that John and Twin John would have had to use it to determine that Twin John used the word *water* to refer to XYZ (Putnam 1975b).

Once language exists, some conventions and prohibitions and definitions can be made explicit, and these explicit rules or commandments or *sensu composito* descriptions can be passed more easily from generation to generation, just as a termite passes on its digestive microbes to the next generation of termites. I believe that this complex descriptive capability is an indispensable part of a modern, human kind of adaptive filtration, an important part of our mechanism for selecting culture itself. Humans make judgments, describe them, and take actions based on the descriptions. What's explicitly mocked, punished, or forbidden may not be repeated as frequently in the future.

If this story about the adaptive functions of language is correct, then the complex, recursive, context-sensitive syntax of human languages makes perfect sense. You would *expect* us, as creatures who use vocal signals in that particular manner, to have much more complex ways of stringing them together than a nightingale or a humpback whale does, because of the much more complicated thing we're doing with them. We're conversing, not simply displaying, encrypting, or transmitting information, even though conversation contains elements of all those. Thought, as we now know it, strikes me as an internalization of conversation, a conversation among fractal elements of the self, among the reflections of reflections in the mirror-maze of our mind.

Somehow the cumulative results of this new human way of doing things produced Socrates and voyages to the moon. Yet how could all that have come out of just smearing our faces with ocher and putting plates in our lower lips and making playful and amusing remarks about these customs as we participated in them? A question I have not yet addressed is how the apparently simple act of domesticating learned behaviors in this way could have produced the insane complexity of a modern human society. Now it's finally time to answer Dennett's (2009a) suggestion that perhaps only technical language is truly domesticated.

CRITICAL MASS

There's one more model of cultural evolution to look at, this time from Enquist and Ghirlanda again, now joined by Arne Jarrick and Carl-Adam Wachtmeister (Enquist et al. 2008). This time, they explored another interesting question: Why has the amount of human culture apparently increased exponentially over long periods of human history? Surveying a variety of sources, they found that the number of significant innovations per year is well described by a simple exponential curve, with annual increases ranging from as high as 3.4 percent (for "Grand Opera" between 1700 and 1900) through a slightly more modest 2.2 percent and 2 percent for genetics and chemistry (1675–1900 and 1550–1900) and a still impressive 1 percent for "important scientific and technical innovations" (1100–1900), all the way down to a glacial .000216 percent per year for "categories of stone tools" between 1.8 million and 225,000 years ago (you see what I mean by long periods of apparent stability).

Looking at European history before 1100, a deep dip in the amount of most sorts of culture would probably show up at some time between the second and eighth centuries. After Tasmania was cut off from the Australian mainland at the end of the last ice age, the complexity of Tasmanian material culture also seems to have collapsed because the remaining inhabitants were too few to maintain it (Henrich 2004). This shows that an exponential increase isn't inevitable, that a society's "carrying capacity" can sometimes crash. But what explains the periods when the increase *is* exponential?

Enquist, Ghirlanda, Jarrick, and Wachtmeister show that the improved transmission of culture from generation to generation can't explain exponential increase, that a less imperfect transmission by itself leads to only a more gradual decline.

Individual creativity on its own, without any story about improved transmission, simply implies that the same skills will be discovered and lost in each generation. A combination of increased individual creativity and more successful transmission produces an asymptotic approach to some fixed limit, a point at which culture is being lost at the same rate it's being created. If the transmission is perfect, the amount of culture will increase endlessly, but at an annual percentage rate that's constantly slowing as the stock being added to grows bigger, unlike the observed exponential increase.

The increase will become exponential only if the amount of new culture produced in a given length of time depends on the amount of culture already existing. That, as the authors point out, is part of the definition of an exponential increase in some x: what x will become next must depend on what x is now.

This means that to produce the observed pattern, the generation of new culture items in each time period, given the existence of a starting amount of culture x, must show some dependency of the form:

$$\gamma(x) = \gamma + \delta x$$

with γ being some underlying invariant rate of generation not dependent on the value of x, $\gamma(x)$ being the overall rate of generation at some value of x, and δx representing the amount by which some amount of existing culture, x, increases the production of new items of culture. If this condition is met, the amount of culture can increase exponentially, at least until it hits some exogenously determined "carrying capacity."

The dynamics, the authors show, is then (once again calling the per-generation loss rate of items of culture λ)

$$\dot{x} = - (\lambda - \delta)x + \gamma$$

If $\lambda > \delta$, this implies that raising δ has the same effect as lowering λ would, leading to stabilization at a somewhat higher level. However, if

$\delta > \lambda$, which can come about as the result of either a rise in δ or a decline in λ (say, as a result of the invention of music or writing), the increase in x becomes exponential, because the absolute change in x over a period of time now depends directly on its starting level. This seems to be the situation we're in today. All other things being equal, you will make more discoveries if you've already invented the epic poem or the university or the telescope or the printing press, than you would have before you invented those things.

Finally, the authors consider the biological evolution of δ, culture-dependent creativity, itself. In this part of the paper, they explicitly assume for simplicity's sake that all culture is adaptive, making it possible to interpret x as the contribution of existing culture to individual fitness. Given this assumption, they show that if there's some fitness cost, $c(\delta)$, which increases as δ increases, selection will favor the increase of δ only if the associated increase in cost, $c(\delta)$, remains less than the corresponding increase in δx.

This obviously depends on x and is logically impossible if it's zero, meaning that the *initial* evolution of culture couldn't have depended on culture-dependent creativity, anymore than it could have depended on improvements in the transmission of culture. There already must be enough culture in existence for new culture to build on, to overcome the cost of a capacity to generate new culture from old. Consequently, the evolution of this capacity must have been a later development. (A development I personally am inclined to associate with the emergence of more modern humans in the past million years, not with the very first steps toward humanity, 2 million or 3 million years ago.) Apparently only an exogenous increase in individual creativity could have begun the whole long process in the first place.

As we've already seen, if we remove the assumption that all culture is adaptive, an ability to recognize when a cultural behavior is no longer working and get rid of it also must have been required from the start. I imagine that the sequence of events must have involved γ and ψ gradually moving up in tandem until the total amount of culture that could be usefully conserved was high enough to begin to produce selection for a higher δ. To continue the speculation to its logical conclusion, the breakthrough point for selection for a higher δ might well have coincided with the transition between Enquist and

Ghirlanda's (2007) two regimes of cultural accumulation, as we moved from a situation in which the amount of culture we could accumulate was limited by our burden of maladaptive culture to a regime in which the accumulation of an unlimited quantity of culture became theoretically possible. If you already possess as much culture as you can carry without poisoning yourself with obsolete and maladaptive practices, the ability to gain more out of the existing stock will be of little use to you. But once you're so good at spotting and weeding out bad versions of behaviors that the accumulation of an unlimited amount of culture truly becomes practical, a self-fueling process of cultural accumulation will be exactly what you need to acquire the skills to displace your less cultivated neighbors. Once you have some common sense, you can afford to be more creative.

This is also the point at which a runaway process of sexual selection for the personal ability to cultivate elaborate forms of culture might have begun, because this is the point at which the disadvantages of an exaggerated version of that ability would have finally disappeared. Before ψ rose high enough to permit unlimited accumulation, too much of the culture that would have been available for a person to accumulate and display would have been actively maladaptive.

The rapid and dramatic character of the "modern revolution" in human behavior, the fact that we moved from the Acheulean hand ax to the orbiting telescope in 500,000 years, after spending considerably more than a million years making essentially the same kind of hand ax, would then be explained by this runaway Fisherian process and by the removal of the obstacle that previously had kept it in check. If true, this would vindicate Darwin's ([1871] 2004) views of the importance of sexual selection in human evolution and reveal his three-way analogy of the lip plate, the fancy pigeon, and the peacock's tail to be extremely telling. Elaborate hairstyles and flattened skulls may not be the most useful things a human being ever invented, but the role of that sort of frivolous adornment in the evolution of the culture-dependent creativity that permitted all our other achievements may in fact have been crucial once we had the common sense needed for unlimited accumulation to become a workable strategy.

Why, mechanically, does already having some culture make the production of even more culture so much easier? Obviously, in the real world, some behaviors and systems of behaviors (for example, writing) must have bigger effects of this kind than others. But Enquist, Ghirlanda, Jarrick, and Wachtmeister cite the general fact that more culture means more individual items that could be made into new combinations with one another. If that sort of permutation or recombination is a common part of the creative process by which new machinery or mathematics or chemistry or opera or philosophy is invented, then invention should become easier as raw material accumulates. This also would explain why it took so long for the whole modern explosion to start. It should be far more difficult to reach critical mass than it is to keep the chain reaction going once you've reached it.

It's possible to construct a simple heuristic model of this process. (Here I'm going outside anything the authors said, so if this part of the story is incorrect, it's my fault.) We can imagine the individual pieces of culture in some domain as nodes on a "graph," which is just a name for a bunch of dots. Those "nodes" may or may not have lines, called "edges," drawn between them. Suppose that for any given node, for each other piece of culture, each other node, there's a constant probability p that a combination of the two would somehow facilitate the production of an additional piece of culture. We can think of these possible "relevances" between items of culture as edges randomly connecting nodes to nodes.

As the absolute number of nodes in the graph, N, rises, the probability for any given node that it's completely irrelevant to everything else, that there's no edge connecting it to any other point in the whole graph, is $(1 - p)^{N-1}$.

So as N goes up, each marginal piece of culture becomes more and more likely to give rise to further combinations, which themselves become more likely to give rise to further combinations. At some level of N that depends on p, there should be a well-defined phase change in which new pieces of culture will go from being rather unlikely to find roles in facilitating combinations to being quite likely to find such roles (Erdös and Rényi [1960] 1976; Kauffman 1969).

Each distinct domain in which innovations are likely to be combined with existing technologies, practices, or ideas eventually will attain recursive critical mass in this way if it accumulates enough items. Every time two items combine to produce a new item of culture, N itself increases by 1, so the process can feed on itself once it begins.

As the absolute amount of our knowledge grows in this way, the individual's epistemic incentive is to put less and less effort into experimental learning, because the rewards of learning what others already know become richer and richer. But to give in completely to this impulse would be to create a world full of people all of whom were knowledgeable, perhaps, but unable to collaborate or even teach because each knew completely different things in a completely different way, making a collective process of careful interpretation impossible. Instead, we moderns borrowed the approach of Euclid's *Elements* or Heron's *Pneumatics*, the great Hellenistic breakthrough in culture-management technology—the explicitly defined, self-consciously conventional technical language, with its own specialized nomothetes and teachers, like the new language devised in Euclid's *Elements* and its many modern descendants.

I've spoken of Dennett's (2009a) apparent suggestion that perhaps only these modern technical languages can truly be thought of as domesticated. In chapter 1, I pointed out that if that's really true, their invention by Hellenistic mathematicians and scientists would have been the original domestication of language, a momentous change in the biology of humans. Instead, I argued that there was a degree of continuity between this way of managing language and its more ordinary or older uses.

Here is Lucio Russo's (2004) description of the basic technology involved in this breakthrough:

> How were the first Greek scientific terms, and therefore the first theoretical entities, created? It is not hard to realize that the essential tools were provided by the postulates of the various theories and the hypothetico-deductive method. Take the first postulate of the *Elements*, for example. In Euclid's text, it reads, literally,
>
> "Let it be demanded that a straight line be drawn (i.e. drawable) from every sign to every sign."
>
> This statement contains words from ordinary Greek, signifying concrete objects: "straight lines" are originally traces drawn or carved (the Greek word γραμμη indicates this explicitly), the "signs" are equally concrete in nature, and the whole can be read as a sentence of ordinary language, with a clear meaning relating to the concrete activity of a draftsman. Naturally the draftsman

can draw, for instance, a green or a red line, thicker or thinner, and make lines of different types. But now suppose that we take this sentence, together with the other four that Euclid wrote, as postulates in his theory based on the hypothetico-deductive method. Because none of the postulates mentions color, clearly no proposition deducible from them can say anything about color. The "lines" of the theory are thus automatically colorless. The same can be said of their thickness or the shape of the "signs." In other words, the use of the hypothetico-deductive method automatically restricts the semantic extension of the terms used in the postulate, generating new entities that are "theoretical" in the sense that the only statements one can make about them are those deducible from the postulates of the theory. (182–83)

It's as if, when learning the new technical language, we had to put on a set of goggles that blinded us to all the qualities of objects except the ones mentioned in the axioms. As geometers, we're color-blind, we no longer see things moving, and if we're doing plane geometry, we can see only two dimensions . . . and all the speakers of the technical language know that all the other competent speakers suffer the exact same induced deficits, all of us know that all of us know that only shape matters and colors don't count in this context, making it easy for us all to attend to the very same aspects of the very same objects in the very same way.

This initial process of abstraction isn't the end of the story, though; it's only its beginning: "Symbols which originally appear to have no meaning whatever acquire gradually, after subjection to what might be called intellectual experimenting, a lucid and precise significance" (Mach [1906] 2004:103). Abstraction combined with repeatable experimentation leads to iterative refinement: "In this career of the similarity notion, starting in its innate phase, developing over the years in the light of accumulated experience, passing then from the intuitive phase into theoretical similarity, and finally disappearing altogether, we have a paradigm of the evolution of unreason into science" (Quine 1969:138).

Because humans, through education and language learning, can be made very different from what they were before their education began, they also can be made very different from one another in some respects, or very similar to one another, or very different from most other humans, and very similar to a *few* others. Learning this new specialized language, which permits us all to describe certain restricted aspects of reality in an unprecedentedly accurate and explicit manner and prevents us all, while we're using it, from collectively even taking note of many others, makes us, in these respects, very different from the rest of humanity and very similar to the other members of our profession or discipline.

This sort of technicalization of a specialized sublanguage, using the method Russo calls "semantic pruning," reducing the number of criteria by which they identify their objects and eliminating many considerations from the descriptions associated with them, along with an explicit focus on logic and rigorous deduction and a discipline-specific obsession with particular tools, or particular instruments of observation, or particular concrete objects of study such as the human body, gave us, in Hellenistic astronomy, mathematics, mechanics, and anatomy, the first really scientific bodies of theoretical knowledge. They were still directly rooted in the physical, visible world, because the terms being semantically pruned were taken directly from ordinary language. (Later Western languages still draw many of their technical terms from classical Greek, so these original acts of abstraction have left their mark on our current language.)

Russo rightly emphasizes the philosophical gap between the essentialism of Plato's and Aristotle's "definitions" and the conventionalism of mathematicians like Euclid and Archimedes a century or two later. This difference looms large for us because it's an important fault line in our own cultural history. Seen from the outside, however, Plato's and Aristotle's first steps down this path are an early and imperfect version of the developing language-management technology, still burdened with unnecessary metaphysical baggage but already moving in the right general direction.

The Hellenistic technology of scientific inquiry rigorously focuses on a single topic, using deliberately invented, formally defined, discipline-specific new vocabulary and creating a precise technical language that will be shared by a specific group of experts, specialized nomothetes for this specific part of our language, who are grouped around a shared set of instruments or methods for performing repeatable experiments or observations. The kind of technical language that Euclid devised is language *as* technology. This seems to connect it in some way to the view that Socrates expressed in *Cratylus*, apparently right around the time the first technical languages of this specific kind were invented, that language, in general, *is* a technology.

Because of the technical language's deliberately circumscribed subject matter, conventions regarding how to think of that subject matter that once were tacit or inadvertent can be made explicit and rationally transparent, and the constant reference back to a well-defined set of concrete objects selects the technical terms for descriptive accuracy, ease of interpretation, and predictive power: "Our language can be seen as an ancient city: a maze of little streets and squares, of old and new houses, and of houses with additions from various periods; and this surrounded by a multitude of new boroughs with straight regular streets and uniform houses" (Wittgenstein 1953:8). The new technical sublanguages are still human language—just as the new modern buildings in Wittgenstein's ancient city still are buildings, not completely different in kind from a sod-roofed hut or a traditional longhouse, just as a Hereford still is a cow and a bloodhound still is a dog—but the process by which they've been shaped into their present form embodies significant innovations in the self-conscious cultivation of better designs. These modern technical languages are managed by self-conscious, entrepreneurial "mavens," peer reviewers and professors and the editors of journals and other specialized nomothetes, in the same way Victorian livestock breeds were managed by self-conscious, entrepreneurial specialized breeders.

We're so used to this way of doing things that it may seem as if it just fell from the sky. But this may not be the proper explanation for what has to be one of the greatest and most productive technological innovations in all of human history. It's rather suspicious that this new technology, the explicitly defined, formal technical language, and the set of institutions needed to maintain it, was discovered in a world in which just a few generations earlier, a theory of meanings that assigns us a self-conscious role in the making and judging of terminology, conceived of as a technology for "teaching and distinguishing natures," had been publicly articulated by its greatest philosopher. Plato was also directly involved in the founding of the Academy, the ultimate institutional ancestor of that central and ubiquitous modern institution for the management, cultivation, and marketing of technical languages, the university.

Even though I'm sure this isn't what happened, it's tempting to imagine Euclid or some predecessor reading *Cratylus* and saying to himself, "Does Socrates really mean that some legislator is supposed to be laying down what we all should mean by the word *parallel*? We'll be waiting a long time for that to happen. Actually, I wonder what I *do* really mean by the word—could it be that I actually mean . . . ?"

The step forward was a huge one, but I don't think that the new way of managing language came out of nowhere, that before Euclid and the birth of the modern technical language, people didn't mean anything in particular by their words, or exercised no care in choosing them, or never directly or indirectly taught one another their meanings. What is Socrates himself doing in the dialogues? The Eleatic Stranger explains the activity in *Statesman*:

> Likenesses which the senses can grasp are available in nature to those real existents which are in themselves easy to understand, so that when someone asks for an account of these existents one has no trouble at all—one can simply indicate the sensible likeness and dispense with the account in words. But to the highest and most important class of existents there are no corresponding visible resemblances, no work of nature clear for all to look on. In these cases nothing visible can be pointed out to satisfy the inquiring mind: the instructor cannot cause the enquirer to perceive something with one or another of his senses and so make him really satisfied that he understands the thing under discussion. Therefore we must train ourselves to give and understand a *rational* account of every existent thing.   (Plato 1961:286e–287a)

To give a rational account of what a thing is, is quite close to giving a rational account of what its name means, and it's partly by giving rational accounts of what they meant by words that Euclid and his contemporaries achieved what they did.

As the complexity of our culture increases and the average degree of each node rises, we have to find ways of encouraging whole communities of people to concentrate on fewer and fewer nodes. If this project has succeeded, if all you really care about is the tau neutrino, it won't help you that the library next door is full of beautiful books about troubadour poetry and human evolution and Hellenistic philosophy. Once you've learned everything that's known about it so far, in order to find out more about the one subject you're obsessively interested in, you will have to resort to experimentation, because at that point, you, like a chimpanzee faced with a new kind of nut, are on your own. Each science seems to require the existence of a community of collaborating experimental learners who have voluntarily put themselves in this position. Each science has its own specialized, highly precise technical language—in which it may be impossible even to mention *Homo ergaster* or Hellenistic philosophers—with its own conventional but amendable definitions. Each science has a set of recognized nomothetes and teachers specifically for that technical language and no other. At a university, each set of specialized nomothetes may be housed in its own special building.

In this sense, modernity is a sort of metaculture with a metalanguage, a network of little groups of somewhat parochial metasavages, each with its own specialized, "semantically pruned" technical vocabulary, each group wearing its own unique set of selectively clarifying and obscuring goggles, and each secretly doubting that the members of the other groups see anything real through their very different goggles. If our tongues are occasionally confused, if the university itself sometimes seems like the Tower of Babel, it's because it is we who have confused them during our many successful but divergent efforts to make various different things clear, each in its own proper way.

The fragmentation of our modern language is simply the institutionalized realization of a possibility implicit in the nature of the human didactic adaptation and of conversation itself. If through education and other forms of conversation, we can be made into much more than we otherwise would have been, we also can be made very different from one another. Like an insect's compound eye, our modern language is a compound language, for a compound culture in which the uniquely human division of epistemic labor has been carried to unprecedented extremes. As Dante ([1303–1305] 1996) observed while describing what ultimately went wrong with the Tower,

> Only among those who were engaged in a particular activity did their language remain unchanged; so, for instance, there was one

for all the architects, one for all the carriers of stones, one for all the stone-breakers, and so on for all the different operations. As many as were the types of work involved in the enterprise, so many were the languages by which the human race was fragmented; and the more skill required for the type of work, the more rudimentary and barbaric the language they now spoke.   (15)

10

Recapitulation and Moral

A BRIEF RECAPITULATION

If I haven't yet convinced you that the hypothesis that the conventions of our languages are a form of domesticated culture is worth entertaining, nothing I can say now is likely to change the situation. If I have succeeded, anything I add here will only detract from that success. The message of my story is right on the surface: that we have more agency in deciding what our language will be like than we might suppose and that we all are working on a shared project that benefits everyone all the time, even when our actions might seem futile or frivolous. So I will finish my narrative by simply reminding you of the overall trajectory of my argument and, for closure, drawing an unsurprising moral at the end.

At the beginning of this book, I introduced the hypothesis, first considered by Daniel Dennett (2009a), that our languages, or the words in them, might be thought of as domesticated. I examined a few of Dennett's reasons for doubting that this hypothesis was true of ordinary language, but I couldn't convince myself that they were genuine obstacles.

I claimed that David Lewis's ([1969] 2002) theoretical model of the conventions of a human language already commits us to a theory of their domestication, at least if we also admit that language is culturally transmitted and that culture can evolve.

To solidify this line of argument, I looked at the model of the language of a human population that Lewis presented in *Convention*, pointing out some of the places where it fit well with my hypothesis. I examined some of the ways that something can become conventional, or cease to be conventional, and found that they all appeared to involve the sort of locally rational human choice that's also involved in the domestication of animals and plants. In the end, however, since Lewis himself never explicitly contemplated the possibility of evolution, I found myself faced by questions that his arguments didn't address.

To consider the specifically evolutionary aspects of human language, I then looked at the theoretical models of the evolution of a closely related, but not identical, kind of communicative conventions articulated by Brian Skyrms (1998, 1999, 2000, 2007, 2009a, 2009b, 2010). This rather different approach seemed to solve some problems. Because of its wider scope, however, it raised a whole new set of questions about the evolution of communication among living things more generally.

In order to deal with these new puzzles, I spent some time thinking about that larger subject. I discussed the nature of biological information in general and identified some of the obstacles to richer forms of communication among nonhuman organisms, culminating with the strange case of the chimpanzee. I suggested that John Maynard Smith and Eörs Szathmáry's (1998) model of the "major transitions" in evolution might provide a good framework for looking at the transition from something like that to ourselves. In the version of the "major transition" story that seemed right to me, this would mean that noisy transmission and internal conflict—conflict between humans and their possibly maladaptive culture—were the key problems that needed to be solved in order to effect the transition, which seemed to create a need for the same kind of "domestication" theory that I had started out with.

To make sure that I wasn't alone in seeing these things as problems, I glanced at some existing models of cultural evolution, in particular Magnus Enquist and Stefano Ghirlanda's (2007; Enquist, Erikson, and Ghirlanda 2007; Enquist et al. 2008) models of "critical learning" and "adaptive filtration." Concluding that these models, too, seemed to support the hypothesis of domestication for culture in general, I next turned to the question of specific mechanisms.

I examined three general, overlapping types of process—the kind of specialized conversation called teaching, initial language acquisition, and, finally, conversation more generally. Education, I argued, is adaptive partly because it's easier for an adult to spot a child's mistake than it is for the child to spot its own. If cultural practices are to be transmitted with enough accuracy to avoid error catastrophe, their transmission apparently requires this sort of active cooperation by both parties involved. I glanced at the impact of this process on human evolution by looking at Kim Sterelny's (2012) "social learning" hypothesis through the lens of a domestication theory of culture.

I then attempted to portray language acquisition as a process of learning to interpret and negotiating one's way into a culture's web of analogies, conducted in the course of drastically simplified conversations in the context of interactions with the immediate shared environment. After briefly explaining Hilary Putnam's (1975b) model of meaning, in which it always depends on the indexical context, and our collaborative efforts to coordinate our picture of that context, I presented the games involved in early language acquisition as drastically simplified and repetitive collaborative contexts that give us opportunities to practice the difficult skill of sharing attention and interpretations, and to develop uniquely human habits like role reversal and recursive mind reading.

I finished by reflecting on the shared human pleasure of laughter and what it tells us about the role of conversation in shaping a group's interpretation of its language and culture. I contended that the behavior we call humor is partly a form of play, partly a form of display, partly a reward for keeping up with the group's consensus reality, and partly a form of adaptive filtration in which interpretations are harmonized or extended by persuasive speakers and become common knowledge in the conversing group. Its role in making the implicit explicit, I argued, also makes it a source of new conventions or a way of clarifying and stabilizing existing ones.

I made an extended comparison between conversation and adaptive immunity, portraying each of them as a fast, cheap way of dealing with the challenge of a rapidly evolving set of opponents. In the case of the immune system, those opponents are our pathogens, but in human evolutionary history, I asserted, the problem has been corrupted or maladaptive items of culture. Finally I closed with some thoughts about the role this set of adaptations must have played in enabling the rise of our very complex modern societies.

The broad contents of the hypothesis that much of human language and human culture—or my version of it—must be thought of as domesticated should now be at least somewhat clearer. I also pointed out that the idea suggests a role for Ronald Fisher's (1930) hypothesis about display in sexual selection in the evolution of our human adaptation for cultivating cumulative culture. This seems to me to raise an interesting question.

The more normal forms of sexual display—singing, bright plumage, acrobatics, pheromones, things like that—have evolved over and over in the history of life in many different kinds of completely unrelated organisms. Is human cultural display the same sort of universal adaptation, something that will emerge again and again in various different forms over the next few hundred million years now that mammals have large enough brains for it, or is it a fluke, something unique and strange, that will never reappear in the history of life on Earth? If the transition to human intelligence and human cumulative culture is just another of Maynard Smith and Szathmáry's "major transitions," is it the sort of transition that will be made again and again on Earth or elsewhere in the universe in more or less the same way as the transition to eusociality has, or is it something that can evolve only once? And if it did evolve again, would the creatures who made the transition the second time also end up with our kind of sense of humor?

Cumulative culture, I think, is too useful to evolve only once. If we ourselves don't kill off all the apes, then I suspect it's an adaptation that they, for one, might evolve over and over, given enough time. Once there were apes in the world, it took tens of millions of years for one lineage to find the adaptation. If we gave the same general kind of organism more tens of millions of years or hundreds of millions of years or a billion, I could easily imagine the same capability evolving several more times.

What isn't as clear to me is whether the details of the way it's achieved would be similar each time. Because a language seems useful for managing cumulative culture, I'm willing to believe that these creatures eventually would evolve a sort of language. The problem of interpretation would inevitably come up if they did, so they would have to have a socialization process in which the young learned how to interpret remarks made using that language. Even as adults, the same general problem would keep coming up and would keep needing to be dealt with, so it wouldn't surprise me to find some sort of acrobatic display of interpretive ability. Sexual selection seems like a plausible mechanism for bringing these capabilities to a human level, although I suppose there could be others. (But there's really nothing like sexual selection for producing florid and exaggerated versions of some feature, so I sometimes wonder whether anything else would work. There may not be any other way of achieving a human kind of intelligence, for the same reason there may not be any other way of creating a peacock's tail.) Depending on the details, you might have a longer or shorter period of making only hand axes and transmitting a sharply limited amount of culture between generations, but again, the unlimited accumulation of culture is so useful that sooner or later, some way of achieving it would probably evolve.

I don't know, however, whether the precise *kind* of interpretation game that we humans play is the only possible one. There might be other, perfectly good substitutes for a sense of humor. Joy in discovery, pleasure at keeping up, and the posing and solving of interpretive puzzles all seem like things we might expect to evolve again and again, but whether that joy would take the form of laughter or some other completely alien response to some other completely alien type of puzzle is a mystery.

The way that humans converse seems to be an exaggerated adaptation of a specific set of cognitive mechanisms that chimpanzees and bonobos already use for play, sex, and toolmaking, coupled with a new set of motives. Another way of doing the same job that came out of a different set of interactions among another kind of animals might look very different from what we do. My best guess is that intelligent orang-utans would have a cognitive adaptation that served the *role* of a sense of humor, but it might scare us, disgust us, or baffle us if we actually saw it. It could easily be mixed up with other behaviors in a way different from humor in humans.

Creatures that evolved from a more distantly related animal might possess something even less recognizable. Bees and termites have solved many of the same evolutionary problems, but the solutions are very different because of the differences in the organism in which the process started. Do dolphins have a sense of humor? Maybe they do or maybe they don't, but it's also possible that they have something else, something that isn't better or worse than a sense of humor, just different, something we'll have a hard time recognizing as part of their particular way of cultivating domesticated culture.

The question of whether dolphins or elephants or sperm whales are "intelligent" in the same way that we are may not have a simple answer. On the one hand, they all lack hands and the resulting possibility of coming up with an evolving material culture. On the other hand, whales and dolphins have sonar, something that we don't have, something that may be susceptible to cultural modifications that are even better than a fancy hairstyle. Their brains are quite big and have been big for much longer than ours have, so they may well be very clever; in some ways, they may even be more capable than we are. But the details are probably so different that a linear comparison would be difficult. Sperm whales, with their huge, complex brains, may be at the very pinnacle and summit of *type 1* intelligence—the type you evolve when you can't make tools or, like the chimpanzee, can't quite sustain a toolmaking tradition—and lack *type 2* intelligence, the kind that we humans have. In some ways, sperm whales may be as superior to us as chimpanzees are to marmosets, just not in the ways that matter most to humans. Over the course of a long life, each individual may become very wise in the ways of the deep ocean without that ever cumulating into any sort of complex cultural tradition. Of course, we don't really know what whales use their big brains for. It's bound to be something very complicated, because everything in nature is complicated when you finally get a clear view of it.

Somewhere in the universe, there probably are other creatures with cognitive adaptations similar to our own, but I now doubt that all intelligence in the universe is arranged on a single linear scale. The story I've told here, with its many accidental and contingent features, suggests that things must be more complex than that.

AN UNSURPRISING MORAL

Because I started this book with *Cratylus* and its implausible suggestions about the origins of language, I'll finish it by reflecting briefly on what must be the single most implausible suggestion in that whole dialogue, something I haven't yet mentioned. I'm speaking of Socrates's claim that a word is supposed to be a kind of sonic portrait of the things it represents, with sounds like lambda representing slick, slimy, liquid kinds of things, and other sounds, like rho, representative of motion, rippling, or rolling, and so on. To be a real, proper, name, he argued, it somehow must be iconic.

Instead, we moderns all believe that the sounds of words are completely arbitrary and conventional. For most purposes, they certainly can be treated that way, but it is true that particular words have particular sounds. We know that some aspects of the way they sound affect their fitness; people tend to avoid words that sound too similar to taboo or obscene words, words that force everyone to think about the fact that everyone's thinking about some unpleasant or upsetting subject. Synonyms with a different sound, or foreign words, are often chosen instead, and the almost-taboo word drops out of the language (Dixon 1997:19).

In a preliterate culture, the words that were preserved and repeated most frequently and in the most attention-grabbing manner were presumably the words used in various kinds of oral culture, including songs, epic poems, and various sorts of stories. The poetic qualities of a word do matter to its employment in poetry. The person who creates or incrementally modifies a song does so on the basis of judgments about the suitability of particular words for conveying the associations and the mood he's trying to evoke. In conversation, we make similar judgments at lightning speed and while our attention is mainly focused on other things. It wouldn't surprise me to find patterns in the relationship between sounds and meanings in a natural language, because that is part of what people think about when they're actually using language.

Isn't this just a needless complication? Yes, it is, and that's the point of the story. Human language is an adaptation of an organism, like the mammalian immune system or the wings of a bird or a bat. When we look closely at such a thing, what we always see is not something clean, simple, and easy to model but an incredible, almost endless wealth of specific, idiosyncratic details, many of which end up mattering a lot. This is a consequence of the nature of the optimization mechanism that produces such systems. It's a consequence of the fact that Nature, the selector, doesn't care how complicated the thing it's being asked to evaluate might be, that like a wind tunnel, Nature can decide in the blink of an eye whether or not something will work, no matter how complex it is. Things can become incrementally more and more complicated, without limit, and so over time, as situations keep coming up in which something can be gained by becoming just a little more complex, they may become very complicated indeed.

The moral with respect to human language is that we should expect it, too, to be very complicated, even more complicated than a fly, although that's already mind-bogglingly, incomprehensibly complicated. We should never expect the full meaning of a word—*justice* or *convention* or *power* or *truth* or *beauty*—to be its dictionary definition and nothing else. Beyond the senses that can easily be described in such a definition is a tangled web of associations and suggestions and implications that is very hard to unravel.

This means that we, as humans, have an impossible responsibility, the responsibility of managing something, our language, that we can never understand completely, something so complex that much of what is important about it is always likely to elude us. We have no choice but to attempt something we're very likely to fail at, through sheer ignorance and stupidity. It matters what we all think justice is, what we all think a convention is, what we all think truth and power and beauty are, but as individuals, we aren't completely competent to decide what any of those words *should* mean.

Fortunately, only a few of our attempts to put our language back into good order need to succeed. It's easier to recognize success than it is to plan it, so the few real successes, like Lewis's ([1969] 2002) success in refining the full set of associations of the word *convention*, may well be recognized as successes over time and accepted into the idiolects of a population of speakers. As individuals, our efforts to improve our language are likely to fail, but in aggregate, we tend to find the right solution sooner or later. That's the problem with evolution; it's incredibly wasteful. A lot of seeds fall on barren ground. A few sprout, but most of them are crowded out by weeds or die in the shadows of taller saplings. Far fewer become trees, but those produce so many seeds that even though most of them fall on barren ground. . . .

This is a daunting prospect, at least for a philosopher, since it means that most of our efforts to clarify things are likely to be futile. Still, it's much better than hearing that we can't possibly succeed at all, ever, no matter what we try, or that the activity itself is unimportant. It takes a lot of courage and perseverance to be the kind of creature we are, but in the end we humans sometimes do achieve spectacular results. It's just that we have to be reckless and daring and try dangerous things, even though we know that most such efforts will fail and that people may legitimately mock our failures.

For me, all this just reinforces, perhaps in a slightly surprising way, a very old set of conclusions about the human role in the world, the conclusions that Longinus, in *On the Sublime*, said that we could draw from Plato's writing style. From the work of that exemplary nomothete, we supposedly learn that "it was not in Nature's plan for us her chosen children to be creatures base and ignoble—no, she brought us into life, and into the whole universe, as into some great field of contest, that we should be at once spectators and ambitious rivals of her mighty deeds, and from the first implanted in our souls an invincible yearning for all that is great, all that is diviner than ourselves" (1890:68).

The phrase in this sentence that now leaps out at me, after all the work that we've done to get to this point, is "spectators and ambitious rivals." We often seem to think of ourselves as mere spectators of nature and even of our own culture and language, passively consuming or imitating what luckily has somehow fallen from the tree. That way of looking at things seems to assume that there is no real difference between a human being and one of Alan Rogers's (1988) imitative snerdwumps. But in his use of language, Plato was more than just a thoughtless imitator. A view of humanity that fails to account for that part of our nature is fatally flawed. In the final analysis, we are not passive spectators, consumers, or victims of our culture. We humans are not puppets dancing at the end of strings held by maladaptive memes—we're gardeners, domesticators and creators, the ambitious rivals of Nature itself.

References

Argiento, R., R. Pemantle, B. Skyrms, and S. Volkov. 2009. "Learning to Signal: Analysis of a Micro-Level Reinforcement Model." *Stochastic Processes and Their Applications* 119: 373–90.

Aristotle. 1952. *The Works of Aristotle*. Edited by W. D. Ross. Vol. 12, *Select Fragments*. Oxford: Oxford University Press.

Aristotle. 1984. *The Complete Works of Aristotle*. Translated by J. Ackrill. Edited by J. Barnes. Vol. 2. Bollingen 71. Princeton, N.J.: Princeton University Press.

Armstrong, D. 2008. "The Gestural Theory of Language Origins." *Sign Language Studies* 8 (3): 289–314.

Aunger, R. 2000. "The Life-History of Culture: Learning in a Face-to-Face Society." *Ethos* 28 (2): 1–38.

Austin, J. L. 1956. "A Plea for Excuses: The Presidential Address." *Proceedings of the Aristotelian Society* 57: 1–30.

Austin, J. L. 1962. *How to Do Things with Words*. Oxford: Clarendon Press.

Axelrod, R. 1984. *The Evolution of Cooperation*. New York: Basic Books.

Axelrod, R., and W. D. Hamilton. 1981. "The Evolution of Cooperation." *Science* 211: 1390–96.

Barham, L. 2002. "Systematic Pigment Use in the Middle Pleistocene of South Central Africa." *Current Anthropology* 43: 181–90.

Baudelaire, C. (1861) 1974. "The Albatross." In *Selected Poems of Charles Baudelaire*, translated by G. Wagner. New York: Grove Press.

Beggs, A. 2005. "On the Convergence of Reinforcement Learning." *Journal of Economic Theory* 122: 1–36.

Binmore, K., and L. Samuelson. 2006. "The Evolution of Focal Points." *Games and Economic Behavior* 55 (1): 21–42.

Boehm, C. 2001. *Hierarchy in the Forest: The Evolution of Egalitarian Behavior.* Cambridge, Mass.: Harvard University Press.

Boesch, C. 1990. "Teaching Among Wild Chimpanzees." *Animal Behaviour* 41: 530–32.

Boesch, C. 1991. "Symbolic Communication in Wild Chimpanzees?" *Human Evolution* 6 (1): 81–89.

Börgers, T., and R. Sarin. 1997. "Learning Through Reinforcement and the Replicator Dynamics." *Journal of Economic Theory* 74: 235–65.

Bowles, S., and H. Gintis. 2012. *A Cooperative Species: Human Reciprocity and Its Evolution.* Princeton, N.J.: Princeton University Press.

Boyd, R., and P. Richerson. 1985. *Culture and the Evolutionary Process.* Chicago: University of Chicago Press.

Boyd, R., and P. Richerson. 1996. "Why Culture Is Common but Cultural Evolution Is Rare." *Proceedings of the British Academy* 88: 77–93.

Boyd, R., and P. Richerson. 2005. *The Origin and Evolution of Cultures.* Oxford: Oxford University Press.

Boyd, R., and J. Silk. 2012. *How Humans Evolved.* 6th ed. New York: Norton.

Bruner, J. 1985. *Child's Talk: Learning to Use Language.* New York: Norton.

Burgess, J. 2013. *Kripke.* Cambridge: Polity.

Burke, G., and M. Strand. 2012. "Polydnaviruses of Parasitic Wasps: Domestication of Viruses to Act as Gene Delivery Vectors." *Insects* 3: 91–119.

Bush, R., and F. Mosteller. 1955. *Stochastic Models of Learning.* New York: Wiley.

Buss, L. 1987. *The Evolution of Individuality.* Princeton, N.J.: Princeton University Press.

Call, J., and M. Carpenter. 2003. "On Imitation in Apes and Children." *Infancia y aprendizaje* 26: 325–49.

Call, J., and M. Tomasello. 1996. "The Effect of Humans on the Cognitive Development of Apes." In *Reaching into Thought: The Minds of Great Apes*, edited by A. Russon, K. Bard, and S. Parker. New York: Cambridge University Press.

Call, J., and M. Tomasello. 2007. "The Gestural Repertoire of Chimpanzees." In *The Gestural Communication of Apes and Monkeys*, edited by J. Call and M. Tomasello. Mahwah, N.J.: Erlbaum.

Carnap, R. 1956. *Meaning and Necessity*. Chicago: University of Chicago Press.

Catchpole, P., and B. Slater. 2008. *Bird Song: Biological Themes and Variations*. Cambridge: Cambridge University Press.

Cavalli-Sforza, L. L., and M. W. Feldman. 1981. *Cultural Transmission and Evolution: A Quantitative Approach*. Princeton, N.J.: Princeton University Press.

Chen, K. H., L. L. Cavalli-Sforza, and M. Feldman. 1982. "A Study of Cultural Transmission in Taiwan." *Human Ecology* 10: 365–82.

Cheney, D. L., and R. M. Seyfarth. 1990. *How Monkeys See the World: Inside the Mind of Another Species*. Chicago: University of Chicago Press.

Chomsky, N. 1965. *Aspects of the Theory of Syntax*. Cambridge, Mass.: MIT Press.

Chomsky, N. (1957) 2002. *Syntactic Structures*. New York: Mouton de Gruyter.

Chomsky, N. 2007. "Of Minds and Language." *Biolinguistics* 1: 9–27.

Cloud, D. 2011. "Biological Information and Natural Selection." In *Information Processing and Biological Systems*, edited by S. Niiranen and A. Ribeiro. Berlin: Springer.

Csibra, G., and G. Gergely. 2006. "Social Learning and Social Cognition: The Case for Pedagogy." In *Processes of Change in Brain and Cognitive Development*, edited by Y. Munataka and M. H. Johnson. Oxford: Oxford University Press.

Csibra, G., and G. Gergely. 2009. "Natural Pedagogy." *Trends in Cognitive Science* 13: 148–53.

Csibra, G., and G. Gergely. 2011. "Natural Pedagogy as Evolutionary Adaptation." *Philosophical Transactions of the Royal Society B* 366: 1149–57.

Custance, D., A. Whiten, and K. Bard. 1995. "Can Young Chimpanzees (*Pan troglodytes*) Imitate Arbitrary Actions? Hayes and Hayes (1952) Revisited." *Behavior* 132 (11–12): 837–89.

Dante. (1303–1305) 1996. *De vulgari eloquentia*. Edited and translated by S. Botterill. Cambridge: Cambridge University Press.

Darwin, C. 1868. *The Variation of Animals and Plants Under Domestication*. 2 vols. London: Murray.

Darwin, C. (1871) 2004. *The Descent of Man and Selection in Relation to Sex*. London: Penguin.

Darwin, C. (1859) 2009. *On the Origin of Species*. London: Penguin.

Dave, A. S., and D. Margoliash. 2000. "Song Replay During Sleep and Computational Rules for Sensorimotor Vocal Learning." *Science* 290 (5492): 812–16.

Davidson, D. 1973. "On the Very Idea of a Conceptual Scheme." *Proceedings and Addresses of the American Philosophical Association* 47: 5–20.

Dawkins, R. 1976. *The Selfish Gene*. Oxford: Oxford University Press.

Dawkins, R. 1982. *The Extended Phenotype: The Long Reach of the Gene*. Oxford: Oxford University Press.

Dawkins, R., and J. Krebs. 1978. "Animal Signals: Information or Manipulation?" In *Behavioral Ecology: An Evolutionary Approach*, edited by J. Krebs and N. Davies. Oxford: Blackwell.

Dawkins, R., and J. Krebs. 1979. "Arms Races Between and Within Species." *Proceedings of the Royal Society of London B* 205 (1161): 489–511.

Dennett, D. 1995. *Darwin's Dangerous Idea: Evolution and the Meanings of Life*. New York: Touchstone.

Dennett, D. 2000. "Making Tools for Thinking." In *Metarepresentations: A Multidisciplinary Perspective*, edited by D. Sperber. Oxford: Oxford University Press.

Dennett, D. 2009a. "The Cultural Evolution of Words and Other Thinking Tools." In *Cold Spring Harbor Symposia on Quantitative Biology*. Cold Spring Harbor, N.Y.: Cold Spring Harbor Laboratory Press.

Dennett, D. 2009b. "The Evolution of Culture." In *Cosmos and Culture: Cultural Evolution in a Cosmic Context*, edited by S. J. Dick and M. L. Lupisella. Washington, D.C.: NASA, Office of External Relations, History Division.

Dixon, R. M. W. 1996. "Origin Legends and Linguistic Relationships." *Oceania* 67 (2): 127–40.

Dixon, R. M. W. 1997. *The Rise and Fall of Languages*. Cambridge: Cambridge University Press.

Doupe, A., and P. Kuhl. 1999. "Birdsong and Human Speech: Common Themes and Mechanisms." *Annual Review of Neuroscience* 22: 567–631.

Dunbar, R. 2003. "The Social Brain: Mind, Language, and Society in Evolutionary Perspective." *Annual Review of Anthropology* 32: 163–81.

Eco, U. 1995. *The Search for the Perfect Language*. Oxford: Blackwell.

Eigen, M. 1971. "Molekulare Selbstorganisation und Evolution." *Naturwissenschaften* 58 (10): 465–523.

Eigen, M. 1992. *Steps Towards Life: A Perspective on Evolution*. Translated by Paul Woolley. Oxford: Oxford University Press.

Eigen, M., M. McCaskill, and P. Schuster. 1988. "Molecular Quasi-Species." *Journal of Physical Chemistry* 92: 6881–91.

Enquist, M., K. Erikson, and S. Ghirlanda. 2007. "Critical Social Learning: A Solution to Rogers' Paradox of Non-Adaptive Culture." *American Anthropologist* 109 (4): 727–34.

Enquist, M., and S. Ghirlanda. 2007. "Evolution of Social Learning Does Not Explain the Origin of Human Cumulative Culture." *Journal of Theoretical Biology* 246 (1): 129–35.

Enquist, M., S. Ghirlanda, A. Jarrick, and C. Wachtmeister. 2008. "Why Does Human Culture Increase Exponentially?" *Theoretical Population Biology* 74 (1): 46–55.

Erdős, P., and A. Rényi. (1960) 1976. "On the Evolution of Random Graphs." In *Selected Papers of Alfréd Rényi*, edited by P. Turan. Vol. 2. Budapest: Akadémiai Kiadó.

Evans, G. 1973. "The Causal Theory of Names." *Aristotelian Society Supplement* 47: 187–208.

Faulhammer, D., A. Cukras, R. Lipton, and L. Landweber. 2000. "Molecular Computation: RNA Solutions to Chess Problems." *PNAS* 97: 1385–95.

Fernald, A., and D. O'Neill. 1993. "Peekaboo Across Cultures: How Mothers and Infants Play with Faces, Voices, and Expectations." In *Parent-Child Play: Descriptions and Implications*, edited by K. McDonald. Albany: State University of New York Press.

Finlayson, C. 2009. *The Humans Who Went Extinct: Why Neanderthals Died Out and We Survived*. New York: Oxford University Press.

Fisher, R. 1930. *The Genetical Theory of Natural Selection*. Oxford: Clarendon Press.

Foley, C., N. Petorelli, and L. Foley. 2008. "Severe Drought and Calf Survival in Elephants." *Biology Letters* 4 (5): 541–44.

Frisch, K. von. 1967. *The Dance Language and Orientation of Bees*. Translated by Leigh E. Chadwick. Cambridge, Mass.: Belknap Press of Harvard University Press.

Fruth, B., and G. Hohmann. 2002. "How Bonobos Handle Hunts and Harvests: Why Share Food." In *Behavioural Diversity in Chimpanzees and Bonobos*, edited by C. Boesch, G. Hohmann, and L. Marchant. Cambridge: Cambridge University Press.

Gardner, M. 1970. "Mathematical Games: The Fantastic Combinations of John Conway's New Solitaire Game 'Life.'" *Scientific American*, October, 120–23.

Gergely, G., and G. Csibra. 2006. "Sylvia's Recipe: The Role of Imitation and Pedagogy in the Transmission of Cultural Knowledge." In *Roots of Human Sociality: Culture, Cognition, Interaction*, edited by N. Enfield and S. Levinson. Oxford: Berg.

Gibbons, A. 2007. "Spear-Wielding Chimps Seen Hunting Bushbabies." *Science* 315: 1063.

Goodman, N. (1955) 1983. *Fact, Fiction, and Forecast*. Cambridge, Mass.: Harvard University Press.

Gould, S., and R. Lewontein. 1979. "The Spandrels of San Marcos and the Panglossian Paradigm: A Critique of the Adaptationist Program." *Proceedings of the Royal Society of London B* 205 (1161): 581–98.

Grice, H. P. 1957. "Meaning." *Philosophical Review* 66 (3): 377–88.

Grice, H. P. 1975. "Logic and Conversation." In *Syntax and Semantics*. Vol. 3, *Speech Acts*, edited by P. Cole and J. L. Morgan. New York: Academic Press.

Grimm, L. 2000. "Apprentice Flint-Knapping: Relating Material Culture and Social Practice in the Upper Paleolithic." In *Children and Material Culture*, edited by J. Safaer-Derevenski. New York: Routledge.

Haagen, C. 1994. *Bush Toys: Aboriginal Children at Play*. Canberra: Aboriginal Studies Press.

Hare, B., A. Rosati, J. Kaminski, J. Braüer, J. Call, and M. Tomasello. 2009. "The Domestication Hypothesis for Dogs' Skills with Human Communication: A Response to Udell et al. (2008) and Wynne et al. (2008)." *Animal Behaviour* 79: e1–e6.

Hare, B., and M. Tomasello. 2005. "Human-Like Social Skills in Dogs?" *Trends in Cognitive Sciences* 9: 439–44.

Hare, B., V. Wobber, and R. Wrangham. 2012. "The Self-Domestication Hypothesis: Evolution of Bonobo Psychology Is Due to Selection Against Aggression." *Animal Behaviour* 83: 573–85.

Harsanyi, J. C., and R. Selten. 1988. *A General Theory of Equilibrium Selection in Games*. Cambridge, Mass.: MIT Press.

Hawkes, K., and N. Blurton Jones. 2005. "Human Age Structures, Paleodemography, and the Grandmother Hypothesis." In *Grandmotherhood: The Evolutionary Significance of the Second Half of Female Life*, edited by E. Voland, A. Chasiotis, and W. Schiefvenhavel. New Brunswick, N.J.: Rutgers University Press.

Hayes, K., and C. Hayes. 1952. "Imitation in a Home-Raised Chimpanzee." *Journal of Comparative Psychology* 45: 450–59.

Henrich, J. 2004. "Demography and Cultural Evolution: How Adaptive Cultural Evolution Can Create Maladaptive Losses: The Tasmanian Case." *American Antiquity* 69 (2): 197–214.

Herman, L., S. Abinchandani, A. Elhajj, E. Herman, J. Sanchez, and A. Pack. 2000. "Dolphins (*Tursiops truncatus*) Comprehend the Referential Character of the Human Pointing Gesture." *Journal of Comparative Psychology* 113 (4): 347–64.

Herrnstein, R. J. 1961. "Relative and Absolute Strength of Response as a Function of Frequency of Reinforcement." *Journal of Experimental Analysis of Behavior* 4: 267–72.

Hewlett, B. S., and L. L. Cavalli-Sforza. 1986. "Cultural Transmission Among Aka Pygmies." *American Anthropologist* 88 (4): 922–34.

Hewlett, B. S., H. Fouts, A. Boyette, and B. Hewlett. 2011. "Social Learning Among Congo Basin Hunter-Gatherers." *Philosophical Transactions of the Royal Society B* 366: 1168–78.

Hickey, D. 1982. "Selfish DNA: A Sexually Transmitted Nuclear Parasite." *Genetics* 101 (3–4): 519–31.

Hohmann, G., and B. Fruth. 2003a. "Culture in Bonobos? Between-Species and Within-Species Variation in Behavior." *Current Anthropology* 44: 563–609.

Hohmann, G., and B. Fruth. 2003b. "Intra- and Inter-Sexual Aggression by Bonobos in the Context of Mating." *Behavior* 140: 1389–1413.

Holzhaider, J., G. Hunt, and R. Gray. 2010. "Social Learning in New Caledonian Crows." *Learning and Behavior* 38 (3): 206–19.

Hoppe, F. M. 1984. "Polya-Like Urns and the Ewens Sampling Formula." *Journal of Mathematical Biology* 20: 91–94.

Horner, V., and A. Whiten. 2005. "Causal Knowledge and Imitation/Emulation Switching in Chimpanzees (*Pan troglodytes*) and Children (*Homo sapiens*)." *Animal Cognition* 8:164–81.

Hrdy, S. 2000. *Mother Nature: Maternal Instincts and How They Shape the Human Species.* New York: Ballantine Books.

Hrdy, S. 2011. *Mothers and Others: The Evolutionary Origins of Mutual Understanding.* Cambridge, Mass.: Belknap Press of Harvard University Press.

Hume, D. (1748) 1993. *An Enquiry Concerning Human Understanding.* Indianapolis: Hackett.

Humphrey, N. 1976. "The Social Function of Intellect." In *Growing Points in Ethology,* edited by P. P. G. Bateson and R. A. Hinde. Cambridge: Cambridge University Press.

Jerne, N. 1984. "The Generative Grammar of the Immune System." Nobel Lecture presented at the Karolinska Institutet, Stockholm, December 8, 1984.

Kandori, M., G. Mailath, and R. Rob. 1993. "Learning, Mutation, and Long Run Equilibrium in Games." *Econometrica* 61: 29–56.

Kauffman, S. 1969. "Metabolic Stability and Epigenesis in Randomly Constructed Genetic Nets." *Journal of Theoretical Biology* 22 (3): 437–67.

Keeley, L., and N. Toth. 1981. "Microwear Polishes on Early Stone Tools from Koobi Fora, Kenya." *Nature* 293: 464–65.

Kegl, J. 1994. "The Nicaraguan Sign Language Project: An Overview." *Signpost* 7 (1): 24–31.

King, S., and V. M. Janik. 2013. "Bottlenose Dolphins Can Use Learned Vocal Labels to Address Each Other." *PNAS* 110 (32): 13216–21.

Kitcher, P. 1985. *Vaulting Ambition: Sociobiology and the Quest for Human Nature.* Cambridge, Mass.: MIT Press.

Kitcher, P. 2011. *The Ethical Project*. Cambridge, Mass.: Harvard University Press.

Kozo-Polyanski, B. (1924) 2010. *Symbiogenesis: A New Principle of Evolution*. Translated by V. Fet and L. Margulis. Cambridge, Mass.: Harvard University Press.

Kripke, S. 1980. *Naming and Necessity*. Cambridge, Mass.: Harvard University Press.

Kripke, S. 1982. *Wittgenstein on Rules and Private Language*. Cambridge, Mass.: Harvard University Press.

Kuhn, T. 1962. *The Structure of Scientific Revolutions*. Chicago: University of Chicago Press.

Kuhn, T. 1990. "The Road Since Structure." *Proceedings of the Biennial Meeting of the Philosophy of Science Association* 2: 3–13.

Lancy, D. 1996. *Playing on Mother Ground: Cultural Routines for Children's Development*. London: Guilford.

Lewis, D. (1973) 2001. *Counterfactuals*. Oxford: Blackwell.

Lewis, D. (1969) 2002. *Convention: A Philosophical Study*. Oxford: Blackwell.

Longinus. (first or third century C.E.) 1890. *On the Sublime*. Translated by A. Lang. London: Macmillan.

Mach, E. (1906) 2004. *Space and Geometry: In the Light of Physiological, Psychological and Physical Inquiry*. New York: Dover.

Malthus, T. (1798) 1983. *An Essay on the Principle of Population*. London: Penguin.

Mann, W. 2000. *The Discovery of Things: Aristotle's Categories and Their Context*. Princeton, N.J.: Princeton University Press.

Margenstern, M. 2013. *Small Universal Cellular Automata in Hyperbolic Spaces: A Collection of Jewels*. Heidelberg: Springer.

Margulis, L. 1970. *Origin of Eukaryotic Cells: Evidence and Research Implications for a Theory of the Origin and Evolution of Microbial, Plant, and Animal Cells on the Precambrian Earth*. New Haven, Conn.: Yale University Press.

Maynard Smith, J. 1970. "Natural Selection and the Concept of a Protein Space." *Nature* 225: 563–64.

Maynard Smith, J. 1978. *The Evolution of Sex*. Cambridge: Cambridge University Press.

Maynard Smith, J. 1982. *Evolution and the Theory of Games*. Cambridge: Cambridge University Press.

Maynard Smith, J., and G. R. Price. 1972. "The Logic of Animal Conflict." *Nature* 246: 15–18.

Maynard Smith, J., and E. Szathmáry. 1998. *The Major Transitions in Evolution*. Oxford: Oxford University Press.

McComb, M., G. Shannon, S. Durant, S. Katito, R. Slotow, J. Poole, and C. Moss.

2011. "Leadership in Elephants: The Adaptive Value of Age." *Proceedings of the Royal Society B* 278 (1722): 3270–76.

McElreath, R., and R. Boyd. 2007. *Mathematical Models of Social Evolution: A Guide for the Perplexed.* Chicago: University of Chicago Press.

Meltzoff, A. N. 1988. "Infant Imitation After a 1-Week Delay: Long-Term Memory for Novel Acts and Multiple Stimuli." *Developmental Psychology* 24: 470–76.

Miller, G. F. 1997. "Sexual Selection for Moral Virtues." *Quarterly Review of Biology* 82 (2): 97–125.

Nelson, D. 1997. "Social Interaction and Sensitive Phases for Birdsong: A Critical Review." In *Social Influences on Vocal Development*, edited by C. Snowdon and M. Hausberger. Cambridge: Cambridge University Press.

Nelson, D., and P. Marler. 1994. "Selection-Based Learning in Birdsong Development." *PNAS* 91 (22): 10498–501.

Olson, M. 1965. *The Logic of Collective Action: Public Goods and the Theory of Groups.* Cambridge, Mass.: Harvard University Press.

Olson, M. 1982. *The Rise and Decline of Nations: Economic Growth, Stagflation, and Social Rigidities.* New Haven, Conn.: Yale University Press.

Pessin, A., and S. Goldberg. 1996. *The Twin Earth Chronicles: Twenty Years of Reflection on Putnam's "The Meaning of 'Meaning.'"* Armonk, N.Y.: Sharpe.

Pierce, N., and E. Winfree. 2002. "Protein Design Is NP-Hard." *Protein Engineering, Design and Selection* 15 (10): 779–82.

Plato. 1961. *The Collected Dialogues of Plato, Including the Letters.* Edited by E. Hamilton and H. Cairns. Bollingen 71. Princeton, N.J.: Princeton University Press.

Plomin, R., J. C. DeFries, G. E. McClearn, and P. McGuffin. 2001. *Behavioral Genetics.* 4th ed. New York: Worth.

Porphyry. (270) 1975. *Isagoge.* Translated by E. W. Warren. Medieval Sources in Translation. Toronto: Pontifical Institute of Medieval Studies.

Potts, R. 1997. *Humanity's Descent: The Consequences of Ecological Instability.* New York: Avon.

Potts, R. 1998. "Variability Selection in Hominid Evolution." *Evolutionary Anthropology* 7 (3): 81–96.

Putnam, H. 1975a. "Language and Philosophy." In *Mind, Language, and Reality.* Vol. 2 of *Philosophical Papers.* Cambridge: Cambridge University Press.

Putnam, H. 1975b. "The Meaning of 'Meaning.'" In *Mind, Language, and Reality.* Vol. 2 of *Philosophical Papers.* Cambridge: Cambridge University Press.

Quine, W. V. 1969. "Natural Kinds." In *Ontological Relativity and Other Essays.* New York: Columbia University Press.

Quine, W. V. (1936) 2004a. "Truth by Convention." In *Quintessence: Basic Readings from the Philosophy of W. V. Quine*, edited by R. F. Gibson. Cambridge, Mass.: Belknap Press.

Quine, W. V. (1951) 2004b. "Two Dogmas of Empiricism." In *Quintessence: Basic Readings from the Philosophy of W. V. Quine*, edited by R. F. Gibson. Cambridge, Mass.: Belknap Press.

Richerson, P., and R. Boyd. 2002. "Institutional Evolution in the Holocene: The Rise of Complex Societies." In *The Origins of Human Social Institutions*, edited by W. G. Runciman. London: British Academy.

Richerson, P., and R. Boyd. 2004. *Not by Genes Alone: How Culture Transformed Human Evolution*. Chicago: University of Chicago Press.

Rizzolatti, G., L. Fadiga, V. Gallese, and L. Fogassi. 1996. "Premotor Cortex and the Recognition of Motor Actions." *Cognitive Brain Research* 3: 2, 131–41.

Roche, H., J.-P. Brugal, A. Delagnes, C. Feibel, S. Harmand, M. Kibunjia, S. Prat, and P.-J. Texier. 2003. "Plio-Pleistocene Archaeological Sites in the Nachukui Formation, West Turkana, Kenya: Synthetic Results 1997–2001." *Comptes rendus palevol* 2 (8): 663–73.

Rogers, A. 1988. "Does Biology Constrain Culture?" *American Anthropologist* 90: 819–31.

Roth, A., and I. Erev. 1995. "Learning in Extensive-Form Games: Experimental Data and Simple Dynamical Models in the Intermediate Term." *Games and Economic Behavior* 8: 164–212.

Rousseau, J.-J. (1781) 1966. "Essay on the Origin of Languages." Translated by J. H. Moran. In *On the Origin of Language*. New York: Ungar.

Russell, B. (1921) 2010. *The Analysis of Mind*. Seaside, Ore.: Watchmaker.

Russo, L. 2004. *The Forgotten Revolution: How Science Was Born in 300 BC and Why It Had to Be Reborn*. Translated by Silvio Levy. Berlin: Springer-Verlag.

Sagan, L. 1967. "On the Origins of Mitosing Cells." *Journal of Theoretical Biology* 14: 225–74.

Savage-Rumbaugh, S., K. McDonald, R. Sevkic, W. Hopkins, and D. Rupert. 1986. "Spontaneous Symbol Acquisition and Communicative Use by Pigmy Chimpanzees (*Pan paniscus*)." *Journal of Experimental Psychology: General* 114: 211–35.

Sbardella, L. 2007. "Philitas of Cos." In *Brill's New Pauly*. Vol. 11, *Antiquity* (*Phi–Prok*), edited by H. Cancik and H. Schneider. Leiden: Brill.

Schelling, T. 1966. *The Strategy of Conflict*. Cambridge, Mass.: Harvard University Press.

"Secrets of the Wild Child." 1997. *Nova*. PBS, March 4.

Seidenberg, M., and S. Petitto. 1987. "Communication, Symbolic Communication, and Language: Comment on Savage-Rumbaugh, McDonald, Sevcik,

Hopkins and Rupert (1986)." *Journal of Experimental Psychology: General* 116 (3): 279–87.

Seyfarth, R., and D. Cheney. 1990. "The Assessment by Vervet Monkeys of Their Own and Another Species' Alarm Calls." *Animal Behaviour* 40 (4): 754–64.

Skyrms, B. 1996. *Evolution of the Social Contract.* Cambridge: Cambridge University Press.

Skyrms, B. 1998. "Salience and Symmetry-Breaking in the Evolution of Convention." *Law and Philosophy* 17: 411–18.

Skyrms, B. 1999. "Stability and Explanatory Significance of Some Simple Evolutionary Models." *Philosophy of Science* 67: 94–113.

Skyrms, B. 2000. "Evolution of Inference." In *Dynamics of Human and Primate Societies: Agent-Based Modeling of Social and Spatial Processes.* Edited by T. Kohler and G. Gumerman. New York: Oxford University Press.

Skyrms, B. 2004. *The Stag Hunt and the Evolution of Social Structure.* Cambridge: Cambridge University Press.

Skyrms, B. 2007. "Dynamic Networks and the Stag Hunt: Some Robustness Considerations." *Biological Theory* 2: 7–9.

Skyrms, B. 2009a. "Evolution of Signaling Systems with Multiple Senders and Receivers." *Philosophical Transactions of the Royal Society B* 364 (1518): 771–79.

Skyrms, B. 2009b. "Presidential Address: Signals." *Philosophy of Science* 75: 489–500.

Skyrms, B. 2010. *Signals: Evolution, Learning, and Information.* Oxford: Oxford University Press.

Slote, M. 1966. "The Theory of Important Criteria." *Journal of Philosophy* 63 (8): 211–24.

Sober, E., and D. Wilson. *Unto Others: The Evolution and Psychology of Unselfish Behavior.* Cambridge, Mass.: Harvard University Press.

Sperber, D. 1996. *Explaining Culture: A Naturalistic Approach.* Oxford: Blackwell.

Sperber, D. 2000. "Metarepresentations in an Evolutionary Perspective." In *Metarepresentations: A Multidisciplinary Perspective*, edited by D. Sperber. Oxford: Oxford University Press.

Sterelny, K. 2012. *The Evolved Apprentice: How Evolution Made Humans Unique.* Cambridge, Mass.: MIT Press.

Stout, D. 2002. "Skill and Cognition in Stone Tool Production: An Ethnographic Case Study from Irian Jaya." *Current Anthropology* 43 (5): 693–722.

Taglialatela, J., J. Russell, J. Schaeffer, and W. Hopkins. 2008. "Communicative Signaling Activates 'Broca's' Homologue in Chimpanzees." *Current Biology* 18: 343–48.

Taylor, P., and L. Jonker. 1978. "Evolutionarily Stable Strategies and Game Dynamics." *Mathematical Biosciences* 40: 145–56.

Thornton, A., and K. McAuliffe. 2006. "Teaching in Wild Meerkats." *Science* 313 (5784): 227–29.

Tomasello, M. 1996. "Do Apes Ape?" In *Social Learning in Animals: The Roots of Culture*, edited by C. Heyes and B. Galef. San Diego, Calif.: Academic Press.

Tomasello, M. 2008. *Origins of Human Communication*. Cambridge, Mass.: MIT Press.

Tomasello, M., J. Call, K. Nagell, R. Olguin, and M. Carpenter. 1994. "The Learning and the Use of Gestural Signals by Young Chimpanzees: A Trans-Generational Study." *Primates* 35: 137–54.

Tomasello, M., J. Call, J. Warren, T. Frost, M. Carpenter, and K. Nagell. 1997. "The Ontogeny of Chimpanzee Gestural Signals: A Comparison Across Groups and Generations." *Evolution of Communication* 1: 223–53.

Tomasello, M., M. Carpenter, and U. Liszowski. 2007. "A New Look at Infant Pointing." *Child Development* 78 (3): 705–22.

Tomasello, M., B. George, A. Kruger, M. Farrar, and A. Evans. 1985. "The Development of Gestural Communication in Young Chimpanzees." *Journal of Human Evolution* 14: 175–86.

Toth, N., K. Schick, S. Savage-Rumbaugh, R. Sevcik, and D. Rumbaugh. 1993. "Pan the Tool-Maker: Investigations into the Stone Tool-Making and Tool-Using Capabilities of a Bonobo (*Pan paniscus*)." *Journal of Archaeological Science* 20 (1): 81–91.

Trut, L. 1999. "Early Canid Domestication: The Farm-Fox Experiment: Foxes Bred for Tamability in a 40-Year Experiment Exhibit Remarkable Transformations That Suggest an Interplay Between Behavioral Genetics and Development." *American Scientist* 87 (2): 160–69.

Wang, H. 1961. "Proving Theorems by Pattern Recognition—II." *Bell System Technical Journal* 40 (1): 1–41.

Wang, H. 1965. "Games, Logic and Computers." *Scientific American*, November, 98–106.

Weissman, A. 1893. *The Germ Plasm: A Theory of Heredity*. Translated by W. Newton Parker and H. Rönnfeldt. New York: Scribner.

West-Eberhard, M. J. 2003. *Developmental Plasticity and Evolution*. Oxford: Oxford University Press.

Westergaard, G., and S. Suomi. 1995. "The Stone Tools of Capuchins (*Cebus apella*)." *International Journal of Primatology* 16 (6): 1017–24.

Whitehead, H. 2003. *Sperm Whales: Social Evolution in the Ocean*. Chicago: University of Chicago Press.

Whiten, A., V. Horner, C. Litchfield, and S. Marshall-Pescini. 2004. "How Do Apes Ape?" *Learning and Behavior* 32 (1): 36–52.

Wilson, E. O. 2012. *The Social Conquest of Earth*. New York: Liveright.

Wittgenstein, L. 1953. *Philosophical Investigations*. Translated and edited by G. E. M. Anscombe. New York: Macmillan.

Wittgenstein, L. (1922) 1998. *Tractatus Logico-Philosophicus*. Translated by C. K. Ogden. New York: Dover.

Wolfram, S. 2002. *A New Kind of Science*. Champaign, Ill.: Wolfram Media.

Wrangham, R. 2009. *Catching Fire: How Cooking Made Us Human*. New York: Basic Books.

Young, H. P. 1993. "The Evolution of Conventions." *Econometrica* 61: 57–84.

Young, H. P. 2001. *Individual Strategy and Social Structure: An Evolutionary Theory of Institutions*. Princeton, N.J.: Princeton University Press.

Index

Abelard, 54
abstraction, 235, 236
Academy (ancient Greek), 237
accumulation (of culture), 137, 156, 157,
    226, 232, 245
Acheulean technology, 89, 157, 226, 227,
    232
Adam (biblical character), 1, 2, 19, 191
adaptation, 23, 24, 35, 95, 97, 104, 110, 121,
    122, 124, 128, 130, 131, 134, 137, 139, 157,
    162, 170, 174, 198, 218, 244–46, 247. *See
    also* didactic adaptation
adaptive filtration, 151–62, 174, 175,
    224–29, 243
adaptive immunity, 224–29, 243
adaptive lag, 147–51, 154
analogies: conspicuous, 69–77, 166, 193; of
    Darwin, 26, 27, 141, 232; of Dennett, 20,
    35; for domestication, 14, 27; as imper-
    fect, 217; Lewis on, 69, 73; natural, 137,

193, 219; as requiring creative inter-
    pretation, 99; role of, 166, 243; salient,
    70, 71, 72, 73, 137, 166, 193; sheepdog as,
    8–9, 201, 224; for vertebrate immune
    system, 227; virus as, 15
analyticity, 189, 190
analytic truth, 5, 37, 39, 40, 41, 191, 217
Anatomy Lesson of Dr. Nicolaes Tulp, The
    (Rembrandt), ii, 8–9, 19, 31, 33
antibodies, 225, 227
ants, 13, 25, 26, 99, 105, 145, 148, 199
apes, 100, 113, 115, 116, 117, 130–31, 132, 139,
    148, 244. *See also* bonobos; chimpan-
    zees; gibbons; orangutans
apprenticeship, 178, 179, 210
Archimedes, 236
argument, as place where culture is man-
    aged, 161
Aristotle, 4, 22, 167, 169, 236
Armstrong, David, 117

artificial selection, 9, 11, 12, 21–22, 25, 26, 27, 35, 173, 204

attentiveness, 71, 72, 73, 134–35

Attic Greek (language), 4

Augustine, Saint, 55

Austin, John L., 6, 9, 195

Australia: Dixon's study of ancient languages of, 111; song lines of ancient, 28

Axelrod, Robert, 108, 109, 110

Babel, 1, 62; Tower of, 239–40

"battle of names," 183–91

Baudelaire, Charles, 211

bees, beehives, 13, 14, 105–6, 245

Beggs, A. W., 87

behaviorism, 7, 53, 81

bilingualism, 30

binding conventions, 63, 65, 66, 68, 136, 137, 180, 181, 189, 190

Binmore, Ken, 70, 134, 172

biological individuals: and beehives, 100–106; clashes among, 128; definition of, 100–101

biological information: beehives and biological individuals, 100–106; birdsong, as syntax without semantics, 106–12; nature of, 35; overview of, 91–100

biological order, 93

birds. See indigobirds; New Caledonian crows; nightingales; parrots; peacocks; songbirds

birdsong, 106–12

blind imitation, 88, 168

bonobos: cognitive mechanisms of, 245; and cooperation, 134, 212; eye contact by, 75; location of, 148; sexuality of, 124, 125, 135; tool use by, 114

Book Reading (game), 194, 195, 196, 199, 201

Börgers, Tilman, 87

Boyd, Robert, 128

brain: of birds, 110, 120; of dolphins, 246; evolution of large, 88, 244; as involved in selection processes, 26; of monkeys, 87–88; role of, 99, 157; of termites, 144–45; of whales, 246

Bruner, Jerome, 192, 193, 194, 195, 197, 199, 200

Burke, Gaelen, 28, 29

Call, Josep, 117, 122

carrying capacity, 157, 229, 230

Categories (Aristotle), 167

Cavalli-Sforza, Luigi Luca, 143, 145, 146, 147

central control, 104

Child's Talk (Bruner), 192

chimpanzees: and additional requirements for adaptive cultural learning, 150; amount of culture of, 152; and assumptions, 211; blind imitation by, 168; cognitive mechanisms of, 245; and cooperation, 212; difference between ancestors of humans and, 158, 179, 193; emulation by, 164, 165–66; and error catastrophes, 97; form of play of, 215; free play of, of selectively inconsequential behaviors, 122–27; gestures of, 62, 215; hitting by, 75; as lacking common interests in performance of shared tasks, 58, 171; as lacking sense of humor, 218; level of complexity of communication of, 100; limitations of, 163; as living in every-person-for-himself world, 76; major transitions of, 127–32; maladaptive directions of, 157–58; and nonnatural meanings, 76; and pointing, 171, 199; and pride, 180; and role-reversal imitation, 198; sclera of eye of, 77; signaling skills of, 150–51; signals of, without syntax, 113–22; social forgetting by, 155; teaching, shared goals, and runaway sexual selection of, 133–42; tool use by, 165; trying to teach how to use fork,

171–72; as unresponsive to opportunities to acquire culture, 144
Chomsky, Noam, 30, 107, 193, 194
coevolutionary arms races, 144
coevolutionary interaction, 225
common ground, 205, 208, 209, 212, 215, 223
common interests, 42–43, 50, 58–59, 119, 171, 204, 206, 207, 209
common knowledge, 46, 48, 49, 50, 51, 170, 171, 174, 178, 179, 180, 184, 206, 208, 209, 210, 211, 243. See also Lewisian common knowledge
complex language, 83, 88, 89, 113, 117, 218
compound language, 239
computational generality, 92
Confessions (Augustine), 55
conspicuous analogies, 69–77, 166, 193
contingent irreversibility, 134
Convention (Lewis), v, 10, 36, 37, 39, 41, 62, 63, 64, 70, 81, 134, 189, 228, 242
conventional implicature, 207, 209
conventions: binding, 63, 65, 66, 68, 136, 137, 180, 181, 189, 190; binding versus nonbinding, 65, 66; contrasted to mannerisms, 29; dynamics of, 51–53; as involving focal points deriving specifically from historical precedent, 48; knowledge of, 53–55; Lewisian, 51, 59, 62, 63, 64, 65, 66, 78, 119, 124, 166, 174, 179, 181, 204, 205; linguistic, 39, 41, 76, 77, 79, 181, 192, 201; meanings of words as, 9; nonbinding, 65, 66, 76, 77, 117, 136; self-interested participation in, 10; skepticism about, 37–42; Skyrmsian, 105; Skyrmsian communicative, 136; Skyrmsian signaling, 78, 79–85, 123, 125; as tools for coordination, 49
conversation: as excludable, 206; Grice's theory of, 203–13; as non-rivalrous, 206; as place where culture is managed, 161; teaching as kind of, 162, 243

conversational implicature, 207
Conway, John, 92
cooperative activity, 133, 160, 212
cooperative culture-filtration equipment, 138
cooperative filtration, 134
cooperative foragers, 177
cooperative games, 42, 43, 62, 63, 66
coordination equilibrium, 45, 63, 66
coordination games, 41, 42–51, 59, 62, 63, 66, 70, 101, 106, 134, 204
coordination problems, 43–44, 47, 48, 49, 50, 69, 159
corruption, 142, 153, 154, 155, 156, 174, 226
Cratylus, 3, 5
Cratylus (Plato), 2, 3, 4, 28, 167, 237, 238, 246
critical learning, models of, 151–62, 242
critical mass, 229–40
Csibra, Gergely, 137, 162, 164, 165, 166, 167, 168, 169, 170, 175
culling processes, 12, 24, 32, 139, 142, 144, 147, 152, 161, 174
cultural accumulation, 137, 156, 157, 226, 232, 245
cultural evolution: amount of, compared with biological evolution, 15; as analogous to domestication, 27; cumulative, 116, 131; as kind of Darwinian evolution, 158; mechanics of, 142; models of, 147, 148, 152, 159, 229, 242; as not always increasing Darwinian fitness, 143; power of, 99; as predating domestication of animals and plants, 24; speed of, 224
"Cultural Evolution of Words and Other Thinking Tools, The" (Dennett), 13
cultural immunity system, 158
cultural knowledge, 168
cultural selection, 99, 162
culture-dependent creativity, 231, 232
cumulative culture, 32, 139, 142, 147, 156, 157, 180, 244

Dante, 239
Darwin, Charles: analogy of, between evolution and domestication, 27; on artificial selection, 9, 11, 12, 21, 26, 35; *The Descent of Man*, 26, 139, 141; *The Origin of Species*, 86, 116; on present as clue to past, 11; on process that gave us flowers, 86; on redundancy, 116–17; on sexual selection, 139, 140, 141, 232; *The Variation of Animals and Plants Under Domestication*, 27; on "war of Nature," 98
Darwinian assumption of continuity, 35
Darwinian domestication, 10, 13, 14, 16, 24, 26, 28, 29, 36, 40, 162
Darwinian evolution, 12, 67, 88, 121, 158
Darwinian fitness, 15, 18, 143, 190
Darwinian interests, 105
Darwinian point of view, 144, 177
Darwinian processes, 68
Darwinians, 34
Darwinian selection, 146
Darwinian terms: benefits, 101; jokes, 218; learning to see behaviors in, 34
Darwinian theoretical alternative, 16
Darwinian veto, 29
Dawkins, Richard, 15
deception, 102–3
declarative knowledge, 56
deliberateness, 82
Democritus, 7, 37, 85
Dennett, Daniel: on alternative of domestication, 15; analogies of, for human relationship with ordinary words, 20; analogy of, with Darwin's model of unconscious artificial selection, 35; on artificial selection, 12; on association with words being maladaptive, 15; "The Cultural Evolution of Words and Other Thinking Tools," 13; on domestication as good model for evolution of technical terms, 9, 229, 234; on domestication as possible solution,

142; on evolution of human language, 13; on evolution of words and music, 11; on languages/words thought of as domesticated, 241; memetic perspective of, 15, 16, 17; on modern technical languages as domesticated, 234; on owning and responsibility for words, 18; on rejection of domestication as theory of evolution of words in ordinary language, 13, 14; on some words as being domesticated, 24; synanthropic model of, of human language, 23, 53; theory of, of difference between ordinary and technical languages, 204
Descent of Man, The (Darwin), 26, 139, 141
description, as process, 188
didactic adaptation, 163–81, 239
disambiguation, 213–24
Discourse on the Origins of Inequality (Rousseau), 63
distinguishing natures, 191. *See also* teaching and distinguishing natures
divide the dollar (game), 44
Dixon, R. M. W., 28, 30, 111, 112
DNA, 92, 93, 130, 133
dogs, 8, 10, 13, 16, 21, 23, 24, 54, 77, 100, 118, 139, 161, 171, 177
dolphins, 77, 115, 116, 246
domestication: Darwinian, 10, 13, 14, 16, 24, 26, 28, 29, 36, 40, 162; definition of, 13; as hypothesis, 11–29; as model for evolution of technical terms, 9, 10, 33–34
Don Quixote (literary character), 224
driving on right-hand side of road example, 48, 49, 101

education, as adaptive, 243
Eigen, Manfred, 94, 95, 96, 130, 142, 146
Elements (Euclid), 4, 18, 234
elephants, 115, 148, 246
emulation, 26, 126, 127, 128, 129, 130, 131, 132, 164–66, 168, 174

encryption, 110, 112, 121, 127
enforcement, mechanisms of, 62, 65
Enquiry Concerning Human Understanding, An (Hume), 168
Enquist, Magnus, 151, 152, 153, 154, 158, 160, 226, 229, 230, 231, 233, 242
environmental instability, 148, 153
equilibrium: coordination, 45, 63, 66; determination of, 48; human language as, 62; Lewisian convention as, 204; multiple possible equilibrium outcomes, 43; non-risk-dominant optimal coordination equilibrium, 66; obvious coordination, 45; payoff-dominant, 64, 71, 135, 172; risk-dominant, 63, 64, 71, 101, 135; signaling, 82; signaling system, 80
Erikson, Kimmo, 151, 152, 158
error catastrophe, 94, 95, 96, 97, 128, 130, 131, 132, 137, 142, 146, 147, 243
error correction, 105, 129, 130, 131, 132, 133, 160
Euclid, 4, 18, 234, 236, 238
Evans, Gareth, 189
evolution: Darwinian, 12, 67, 88, 121, 158; of human language, 4, 13, 24, 35, 89, 108, 111, 124, 183; of language, 11, 35, 161, 190; of large brain, 88, 244; major transitions in, 116, 209, 242, 244; of semantics, 29–36, 89, 213; of signals, 85–90; of syntax, 29, 31, 89; of technical terms, 9, 10, 33–34; as theory of origin of modern technical languages, 18; as wasteful, 248. *See also* cultural evolution; human evolution
evolutionary arms race, 140, 143–47, 225
"Evolution of Conventions, The" (Young), 9, 64, 80
evolved apprentice, 175–81
experimental learning/experimentalism, 126, 149, 150, 151, 173, 174, 234, 239
explicit bargaining, 160
explicitness, 82

eye, sclera of: chimpanzee, 77; human, 77, 118

Faulhammer, Dirk, 107
Feldman, Marcus, 143, 147
Fernald, Anne, 194
filtration: adaptive, 151–62, 174, 175, 224–29, 243; cooperative, 134; cooperative culture-filtration equipment, 138; intelligent, 32, 152
first legislators, as creating basic elements of human languages, 2, 3
Fisher, Ronald, 141, 244
Fisherian process, 232
flouting, 214, 217
focal points, 41, 47, 48, 159, 180
Forgotten Revolution, The (Russo), 4
free play of selectively inconsequential behaviors, 122–27

games: Book Reading, 199, 201; classes of, 42; cooperative, 42, 43, 62, 63, 66; coordination, 41, 42–51, 59, 62, 63, 66, 70, 101, 106, 134, 204; divide the dollar, 44; Game of Life, 92; impure coordination, 43, 44; language, 192, 194, 201; noncooperative, 42; peekaboo, 194, 195, 196, 197–99, 200, 201; prisoner's dilemma, 108–9; pure coordination, 43, 44, 101; zero-sum, 42
game theory, 9, 70, 134, 176
Gergely, György, 137, 162, 164, 165, 166, 167, 168, 169, 170, 175
"Gestural Repertoire of Chimpanzees, The" (Call and Tomasello), 117
Ghirlanda, Stefano, 151, 152, 153, 154, 158, 160, 226, 229, 230, 231, 232, 233, 242
gibbons, 89, 90, 115
Goldberg, Sanford, 184
Goodall, Jane, 124
Goodman, Nelson, 23
Gould, Stephen Jay, 123
grammar, 30, 33, 69, 110, 111, 193, 195, 227

grandmothers, importance of in human evolution, 139
Greek (language), 4, 34, 234, 236
Grice, H. P., 72, 73, 75, 85, 204, 206, 207, 209, 210, 213, 214, 215; "Logic and Conversation," 204
Gricean attributions, 82
Gricean behavior, 77
Gricean elements, 200
Gricean gesture, 199
Gricean indicative pointing, 173
Gricean kind of attention management, 77
Gricean nonnatural meaning, 73, 74, 76, 78
Gricean spirit, 169
Gricean thwack on the head, 177

Hamilton, William, 108, 109, 110
Harsanyi, John, 64
Henrich, Joseph, 152
Hermogenes, 2
Heron, 234
Hewlitt, Barry, 146
higher-order expectations, 47, 65, 79, 119
historical-chain-of-transmission theory of reference, 9
Homer, 3, 4
Homeric Greek (language), 4, 34
human evolution: importance of grandmothers in, 139; importance of sexual selection in, 140, 232; Machiavellian perspective on, 176; macrotheory of, 175; and "social learning" hypothesis, 176, 243; theories of, 88, 139; understanding of, 98
human knowledge, sources of, 168
human language: as adaptation, 247; chimpanzee gestures compared with, 117, 121; as complex, 7, 111, 113; conventions of, 38–77; Dennett's synanthropic model of, 23, 53; evolution of, 4, 13, 24, 35, 89, 108, 111, 124,

183; evolutionary aspects of, 242; first legislators as creating basic elements of, 2, 3; model of conventions of, 242; moral with respect to, 248; natural, 23, 34; origins of, 85, 105; syntax of, 225, 228; telling truth in, 60. See also language
Hume, David, 7, 168, 169
hummingbirds, 106, 223
humor, 213–24, 243, 244, 245, 246. See also jokes; laughter

idiolects, 12, 18, 41, 67, 68, 96, 130, 190, 192, 199, 219, 222, 248
Iliad (Homer), 28
imitation, 32, 38, 81, 82, 120–21, 126, 127, 128, 130, 131, 132, 146, 147–51, 155, 163, 164, 166, 168, 173, 223. See also blind imitation; role-reversal imitation
implicature, 206, 207, 209, 213, 216, 219
impure coordination game, 43, 44
indicative pointing, 76, 77, 119, 171, 173, 199
indigobirds, 106, 110
information, in cells, 94. See also biological information
information bottleneck, 83, 87
information processing, 90, 91, 92, 97, 98
informative pointing, 171
inheritance, 10, 22, 128, 130, 138, 146, 156, 224
insects, 106, 127, 145. See also ants; bees; beehives; termites
instruction and rehearsal, 201
intelligence, type 1 and type 2, 246
intelligent design, 15, 16
intelligent filtration, 32, 152
intention movement, 118
interpret, learning to, 192–201
interpretation, 56, 57, 69, 73, 74, 110, 188, 190, 191, 194, 196, 213–16, 218–24, 234, 243, 244–45
interpretive puzzles, 215, 216, 223, 245

Jarrick, Arne, 229, 230, 233
joke-in-conversation, 221
jokes, 74, 204, 214–24
Jonathan (research subject), 197–99
Jonker, Leo, 82

Kandori, Michihiro, 63, 64
knowledge: cultural, 168; decorative,
    56; human, sources of, 168; private,
    209; *sensu composito*, 54, 55, 228;
    sensu diviso, 54, 55. See also common
    knowledge
Kripke, Saul, 9, 18, 186, 194

language: acquisition of, 181, 192, 193, 194,
    195, 243; complex, 83, 88, 89, 113, 117,
    218; compound, 239; evolution of, 11,
    35, 161, 190; Greek, 4, 34, 234, 236; of a
    human population, 56–68; Lewisian,
    74; Nicaraguan Sign Language, 28;
    pidgin, 30, 110; prestige, 112; role of, in
    elimination of maladaptive culture,
    159; signaling, 68, 72, 82, 84, 115; tech-
    nical, 3, 9, 13, 17, 18, 38, 201, 204, 229,
    234–39. See also human language
language games, 192, 194, 201
language instinct, 30
laughter, 216, 217, 218, 220, 221, 222, 223,
    243, 245
learning by reinforcement, 82
legal definitions, 4
Levallois technique, 18
Lewis, David: on analogies, 69, 93; on any
    particular language being the language
    of any human population, 67; on
    being trapped by convention, 59–60;
    on clashes of idiolects, 222; on com-
    mon interests, 58–59; on complicated
    events, 99; *Convention*, v, 10, 36, 37, 39,
    41, 62, 63, 64, 70, 81, 134, 189, 228, 242;
    on conventions becoming traps, 154;
    on conventions coming and going, 53;
    on conventions managing to remain

stable, 79, 80, 159; on coordinating
    around new convention, 160; on
    coordination games, 43; definition of
    convention of, 48, 85, 153; driving on
    right-hand side of road example, 48,
    49, 101; on explicitness, 82; on human
    behavior, 62; on idiolects, 96; on
    knowledge of conventions, 54, 55; on
    logical notation, 67; on mannerisms,
    29; model of conventions of human
    language of, 242; model of signaling
    system of, 80; on nonnatural mean-
    ing, 74–75; on questioning Russell's
    rejection of public linguistic conven-
    tions, 41; on refining associations of
    convention, 248; on regularity ceasing
    to be conventional, 51; on sense of
    word *convention*, 10; on signals, 72;
    on Skyrmsian conventions, 81; on
    social contract as compared with
    convention, 60, 61; on truth condi-
    tions of imperative sentence, 58; on
    verbal signaling languages/system,
    56–57, 82
Lewisian binding convention, 66
Lewisian common knowledge, 51, 52, 70,
    81, 172, 174, 200, 207
Lewisian conventions, 51, 59, 62, 63, 64,
    65, 66, 78, 119, 124, 166, 174, 179, 181,
    204, 205
Lewisian higher-order expectations, 47
Lewisian languages, 74
Lewisian recursive replication of mental
    states, 172
Lewisian signaling language, 115
Lewisian signals, 74
Lewisian social conventions, 59
Lewisian theoretical framework, 29–30
Lewontein, Richard, 123
lexicographers, 1, 3, 14, 40, 191
lies, discernment of, 101–3
linguistic conventions, 39, 41, 76, 77, 79,
    181, 192, 201

"Logic and Conversation" (Grice), 204
Longinus, 249

Mailath, George, 63, 64
major transitions, 104, 116, 127–32, 138,
    209, 242, 244
Major Transitions in Evolution, The (Smith
    and Szathmáry), 103
maladaptive culture, 129, 132, 133, 137, 142,
    145, 146, 147, 154, 156, 157, 159, 160, 225,
    226, 232, 242
malapropisms, 217
mannerisms, 29, 30, 81, 110
master sequence, 96
mathematical definitions, 4
Maynard Smith, John, 103, 104, 105, 116,
    127, 134, 142, 209, 242, 244
meaning, model of, 243
"Meaning of 'Meaning,' The" (Putnam),
    184, 187
meanings, of words, 9, 10, 183–202. *See
    also* natural meaning; nonnatural
    meaning
meme, 15, 249
memetic approach, 17, 24
memetic perspective, 15, 16, 17, 24
Mendelian transmission, 146
Metaphysics (Aristotle), 4
mirror neurons, 120
multiple possible equilibrium outcomes,
    43
mutations, 12, 33, 94, 95–97, 129, 130, 131,
    133, 134, 160, 161, 224
mutualism, 13, 14, 20, 26, 28, 29, 144

"names, battle of," 183–91
Naming and Necessity (Kripke), 186
natural analogies, 137, 193, 219
natural meaning, 72
natural selection, 12, 93, 94, 97, 98–99, 104,
    130, 156, 161
neutral framework, 12
New Caledonian crows, 178

Nicaraguan Sign Language, 28
nightingales, 89, 106, 107–8, 110, 114, 121,
    140, 223, 228
noise, 23, 65, 92, 94, 96, 129, 137, 138, 139,
    142, 146, 188, 211, 226
noisy transmission, 41, 147, 242
nomothetes, 2, 3, 5, 6, 41, 219, 224, 234, 236,
    237, 239, 249
nonbinding conventions, 65, 66, 76, 77,
    117, 136
noncooperative games, 42
nonnatural meaning, 72, 73, 74, 75, 76,
    78, 85
non-risk-dominant optimal coordination
    equilibrium, 66

obligatory sexual reproduction, 104, 105
obvious coordination equilibrium, 45
obviousness, 44, 45, 46, 47, 48, 75, 220
Oldowan technology, 164
one if by land, two if by sea example, 50,
    51, 57, 68, 72, 80, 83, 85, 86, 124, 136, 166
O'Neill, Daniela, 194
On Philosophy (Aristotle), 22
On the Sublime (Longinus), 249
ontogenic ritualization, 120, 125
optimization mechanism, 152, 247
orangutans, 100, 245
original legislators, 6
Origin of Species, The (Darwin), 86, 116

paradox: of Plato, 5, 6, 7, 41, 82; of Quine,
    1, 4, 6, 7, 41
parasites, parasitism, 21, 99, 104, 106, 110,
    129, 134, 137, 138, 139, 142, 144, 146, 154,
    158, 161, 225, 227
parents, role of, 71, 135, 145, 146, 180, 181,
    195, 196
parrots, 106, 110, 120, 126, 150
pathogens, 142, 144, 158, 225, 227, 244
Pavlovian reinforcement, 87, 88
payoff-dominant amount of attention,
    71, 134

payoff-dominant equilibrium, 64, 71, 135, 172

peacocks, 26, 27, 103, 112, 114, 232, 245

peekaboo (game), 194, 195, 196, 197–99, 200, 201

Pessin, Andrew, 184

Philitas of Cos, 3

Philosophical Investigations (Wittgenstein), 192

pidgin languages, 30, 110

Plato, 2, 3, 4, 5, 6, 7, 20, 28, 41, 69, 82, 167, 224, 236, 237, 238, 249

Platonism, 191

Pneumatics (Heron), 234

pointing, 75, 118, 199, 200. *See also* indicative pointing; informative pointing

political satire, 220

Pólya-urn process, 84

precedent, 41, 46, 48, 50, 53, 62, 65, 75, 119, 159, 174

prestige languages, 112

Principia Mathematica (Whitehead and Russell), 53

principle of quality, 214

prisoner's dilemma (game), 108–9

private knowledge, 209

pure coordination game, 43, 44, 59, 101

Putnam, Hilary, 18, 63, 101, 184, 185, 186–87, 189, 194, 196, 243

quality, principle of, 214

quasispecies, 96

Quine, W. V., v, 2, 3, 4, 5, 6, 7, 31, 37, 38, 39–40, 41, 55, 81, 82, 189, 191

rational choice, 8, 16, 17, 23, 42, 56, 152

rational transparency, 34

recombination: sexual, 94, 97, 130; V(J)D, 225

recursive mind-reading, 200

recursive tasks, 163–75

redundancy, 116, 117

reinforcement: learning by, 82; models of, 87, 89, 90, 123; Pavlovian, 87, 88

Rembrandt van Rijn, v, 8, 33

replicator dynamics, 82, 87

Richard (research subject), 200–201

Richerson, Peter, 128

risk-dominant amount of attention, 71, 134, 172

risk-dominant equilibrium, 63, 64, 71, 101, 135

RNA, 92, 93, 97, 107, 225

Rob, Rafael, 63, 64

Rogers, Alan, 147, 148, 150, 151, 152, 249

role reversal, 198, 200, 243

role-reversal imitation, 120, 198

Rousseau, Jean-Jacques, 63, 153, 154, 176

rule givers, 5

Russell, Bertrand, 37–39, 41, 53, 81

Russo, Lucio, 4, 234, 236

Samuelson, Larry, 70, 134, 172

Sancho Panza (literary character), 224

Sarin, Rajiv, 87

Schelling, Thomas, 41, 43, 44, 45, 46, 48, 62, 79, 160, 180

scientific inquiry, 236

selection event, 12, 190

self-domestication, 124

Selten, Reinhard, 64

semantic pruning, 236, 239

semantics: definition of, 29; evolution of, 29–36, 89, 213; syntax without (birdsong), 106–12

sensu composito knowledge, 54, 55, 228

sensu diviso knowledge, 54, 55

sexual recombination, 94, 97, 130

sexual selection, 26, 27, 32, 104, 139–41, 144, 232, 244–45

"sexy sons" hypothesis, 141

sheepdog analogy, 8–9, 201, 224

signaling conventions, Skyrmsian, 78, 79–85, 123, 125

signaling equilibrium, 82

signaling languages, 68, 72, 82, 84, 115
signaling system equilibriums, 80
signaling systems, 50, 56, 59, 80, 81, 83, 86, 89, 103, 106, 112, 116, 125
signals, evolution of, 85–90
Signals (Skyrms), 36, 77, 85
Skyrms, Brian: on deliberateness, 82; on evolution of signaling conventions, 9; on information bottleneck, 83; on learning through reinforcement, 82; models of conventions of human language of, 242; models of reinforcement of, 87, 89, 90, 123; on origins of language, 7; on Pólya-urn process, 84; semiaccidental method of, 122; on signaling conventions, 80, 81; *Signals*, 36, 77, 85
Skyrmsian communicative conventions, 136
Skyrmsian conventions, 105
Skyrmsian puzzle, 113
Skyrmsian signaling conventions, 78, 79–85, 123, 125
Skyrmsian signaling language, 115
Skyrmsian signaling system, 116
Skyrmsian signals, 101, 102, 105, 106, 108, 114, 116, 121, 122
snerdwumps, 148, 150, 151, 249
social contracts, v, 31, 38, 59, 60, 61, 62, 65, 72, 73, 124, 136, 172, 177, 189, 215, 219, 221
"social intelligence" hypothesis, 176
"social learning" hypothesis, 162, 176, 243
"social-pragmatic" theory of language acquisition, 192
Socrates, 2, 3, 4, 5, 18, 180, 191, 224, 229, 237, 238, 247
Socratic selection, 33
Socratic story, 2
Solon, 4, 6
songbirds, 106–12, 114, 121, 209
Sophist (Plato), 167
sounds, of words, 247
spandrel, 123

speech acts, 170, 195
Sperber, Daniel, 75, 77
Spooner, Reverend, 216
stag hunt example, 63, 64, 101, 153, 154, 159, 161
Statesman (Plato), 167, 238
Sterelny, Kim, 128, 133, 134, 135, 138, 162, 175–76, 177, 178, 179, 212, 243
Stewart, Potter, 55
Strand, Michael, 28, 29
symbiogenesis, 104, 138
synanthropic model, 53
syndics, v, 1, 3, 5, 6, 8, 31, 41
syntactic mannerisms, 29
syntax: evolution of, 29, 31, 89; learning of, 194; as more or less arbitrary, 193; signals without, 113–22; without semantics (birdsong), 106–12
Szathmáry, Eörs, 103, 104, 105, 116, 127, 134, 142, 209, 242, 244

tacit bargaining, 44, 160
tags, meanings of words as, 9
Tasmania, material culture of, 152, 229
Taylor, Peter, 82
teachers, role of, 1, 3, 6, 53, 71, 72, 133, 135, 137, 139, 161, 167, 168, 174–75, 180
teaching: as form of domestication of culture, 137; as kind of conversation, 162; psychological adaptations for, 137; role of teachers in, 1, 3, 6, 53, 71, 72, 133, 135, 137, 139, 161, 167, 168, 174–75, 180; as specialized conversation, 243
teaching and distinguishing natures, 3, 20, 167, 180, 237
technical terms/language, 3, 9, 13, 17, 18, 38, 201, 204, 229, 234–39
technology, 3, 17, 176–78, 234, 236–37. *See also* Acheulean technology; Levallois technique; Oldowan technology; tool use
teleological opacity, 166, 170
termites, 24, 144–45, 228, 245

Tomasello, Michael, 58, 75–76, 77, 117, 119, 122, 138, 192, 193, 208, 212

tool use, 24–25, 76, 114, 125–27, 128, 132, 133, 155, 164, 165, 175, 178, 179, 180, 181, 212, 229, 245, 246. *See also* Acheulean technology; Levallois technique; Oldowan technology

Tractatus Logico-Philosophicus (Wittgenstein), 6

trial and error, 82, 132, 150, 151

trophallaxis, 144, 145

true by definition, 5, 39, 40, 182, 189, 190, 191

"Truth by Convention" (Quine), 37, 39

Tulp, Dr., ii, 8–9, 19, 31, 161, 182, 222

Turing machine, 91, 106, 107

Twin Earth Chronicles (Pessin and Goldberg): Earth and Twin Earth in, 185, 187; John and Twin John in, 184–85, 187, 202, 228;

"Two Dogmas of Empiricism" (Quine), 39, 40

unconscious artificial selection, 11, 35, 173

unsafe mine example, 59, 152, 154, 159

Variation of Animals and Plants Under Domestication, The (Darwin), 27

verbal signaling system, 56

vertical transmission, 146, 147

vervet monkeys, 86, 87–88, 103

virus analogy, 15

viruses, 14, 15, 28, 101, 144, 158, 190, 226

V(J)D recombination, 225

Wachtmeister, Carl-Adam, 229, 230, 233

Wallace, Alfred Russel, 140

Wang, Hao, 91

Wang tiles, 91, 92

Weissman, August, 100

welfare of words, 20

whales, 89, 228, 246

wit, display of, 214, 217

Wittgenstein, Ludwig, 6, 23, 192, 237

Wolfram, Stephen, 91

wolves, 100, 161, 171

Young, H. Peyton, 9, 63, 64, 65, 66, 77, 80

Youngian nonbinding convention, 66, 77

zero-sum games, 42